PRACTICAL
INDUCTIVELY
COUPLED PLASMA
SPECTROSCOPY

Analytical Techniques in the Sciences (AnTS)

Series Editor: David J. Ando, Consultant, Dartford, Kent, UK

A series of open learning/distance learning books which covers all of the major analytical techniques and their application in the most important areas of physical, life and materials sciences.

Titles Available in the Series

Analytical Instrumentation: Performance Characteristics and Quality
Graham Currell, University of the West of England, Bristol, UK

Fundamentals of Electroanalytical Chemistry
Paul M.S. Monk, Manchester Metropolitan University, Manchester, UK

Introduction to Environmental Analysis
Roger N. Reeve, University of Sunderland, UK

Polymer Analysis
Barbara H. Stuart, University of Technology, Sydney, Australia

Chemical Sensors and Biosensors
Brian R. Eggins, University of Ulster at Jordanstown, Northern Ireland, UK

Methods for Environmental Trace Analysis
John R. Dean, Northumbria University, Newcastle, UK

Liquid Chromatography–Mass Spectrometry: An Introduction
Robert E. Ardrey, University of Huddersfield, Huddersfield, UK

Analysis of Controlled Substances
Michael D. Cole, Anglia Polytechnic University, Cambridge, UK

Infrared Spectroscopy: Fundamentals and Applications
Barbara H. Stuart, University of Technology, Sydney, Australia

Practical Inductively Coupled Plasma Spectroscopy
John R. Dean, Northumbria University, Newcastle, UK

Forthcoming Titles

Techniques of Modern Organic Mass Spectroscopy
Robert E. Ardrey, University of Huddersfield, Huddersfield, UK

Analytical Profiling and Comparison of Drugs of Abuse
Michael D. Cole, Anglia Polytechnic University, Cambridge, UK

Quality Assurance in Analytical Chemistry
Elizabeth Prichard and Vicki Barwick, Laboratory of the Government Chemist, Teddington, UK

PRACTICAL INDUCTIVELY COUPLED PLASMA SPECTROSCOPY

John R. Dean
Northumbria University, Newcastle, UK

John Wiley & Sons, Ltd

Other Wiley Editorial Offices

John Wiley & Sons Inc., 111 River Street, Hoboken, NJ 07030, USA

Jossey-Bass, 989 Market Street, San Francisco, CA 94103-1741, USA

Wiley-VCH Verlag GmbH, Boschstr. 12, D-69469 Weinheim, Germany

John Wiley & Sons Australia Ltd, 42 McDougall Street, Milton, Queensland 4064, Australia

John Wiley & Sons (Asia) Pte Ltd, 2 Clementi Loop #02-01, Jin Xing Distripark, Singapore 129809

John Wiley & Sons Canada Ltd, 22 Worcester Road, Etobicoke, Ontario, Canada M9W 1L1

Wiley also publishes its books in a variety of electronic formats. Some content that appears in print may not be available in electronic books.

Library of Congress Cataloging-in-Publication Data:

Dean, John R.
  Practical inductively coupled plasma spectroscopy / John R. Dean.
     p. cm. – (Analytical techniques in the sciences)
  Includes bibliographical references and index.
  ISBN 0-470-09348-X (cloth : alk. paper) – ISBN 0-470-09349-8 (pbk. :
alk. paper)
 1. Inductively coupled plasma spectrometry. I. Title. II. Series.
  QD96.I47D43 2005
  543'.65–dc22

                                   2005004183

British Library Cataloguing in Publication Data

A catalogue record for this book is available from the British Library

ISBN-13 978-0-470-09348-1 (Cloth) 978-0-470-09349-8 (Paper)
ISBN-10 0-470-09348-X (Cloth) 0-470-09349-8 (Paper)

Typeset in 10/12pt Times by Laserwords Private Limited, Chennai, India

This book is printed on acid-free paper responsibly manufactured from sustainable forestry in which at least two trees are planted for each one used for paper production.

To Lynne, Sam and Naomi for all their patience

Contents

Series Preface

There has been a rapid expansion in the provision of further education in recent years, which has brought with it the need to provide more flexible methods of teaching in order to satisfy the requirements of an increasingly more diverse type of student. In this respect, the *open learning* approach has proved to be a valuable and effective teaching method, in particular for those students who for a variety of reasons cannot pursue full-time traditional courses. As a result, John Wiley & Sons, Ltd first published the Analytical Chemistry by Open Learning (ACOL) series of textbooks in the late 1980s. This series, which covers all of the major analytical techniques, rapidly established itself as a valuable teaching resource, providing a convenient and flexible means of studying for those people who, on account of their individual circumstances, were not able to take advantage of more conventional methods of education in this particular subject area.

Following upon the success of the ACOL series, which by its very name is predominately concerned with Analytical *Chemistry*, the *Analytical Techniques in the Sciences* (AnTS) series of open learning texts has been introduced with the aim of providing a broader coverage of the many areas of science in which analytical techniques and methods are increasingly applied. With this in mind, the AnTS series of texts seeks to provide a range of books which covers not only the actual techniques themselves, but *also* those scientific disciplines which have a necessary requirement for analytical characterization methods.

Analytical instrumentation continues to increase in sophistication, and as a consequence, the range of materials that can now be almost routinely analysed has increased accordingly. Books in this series which are concerned with the *techniques* themselves reflect such advances in analytical instrumentation, while at the same time providing full and detailed discussions of the fundamental concepts and theories of the particular analytical method being considered. Such books cover a variety of techniques, including general instrumental analysis, spectroscopy, chromatography, electrophoresis, tandem techniques, electroanalytical methods, X-ray analysis and other significant topics. In addition, books in this series also consider the *application* of analytical techniques in areas such as environmental science, the life sciences, clinical analysis, food science, forensic analysis, pharmaceutical science, conservation and archaeology, polymer science and general solid-state materials science.

Written by experts in their own particular fields, the books are presented in an easy-to-read, user-friendly style, with each chapter including both learning objectives and summaries of the subject matter being covered. The progress of the reader can be assessed by the use of frequent self-assessment questions (SAQs) and discussion questions (DQs), along with their corresponding reinforcing or remedial responses, which appear regularly throughout the texts. The books are thus eminently suitable both for self-study applications and for forming the basis of industrial company in-house training schemes. Each text also contains a large amount of supplementary material, including bibliographies, lists of acronyms and abbreviations, and tables of SI Units and important physical constants, plus where appropriate, glossaries and references to literature and other sources.

It is therefore hoped that this present series of textbooks will prove to be a useful and valuable source of teaching material, both for individual students and for teachers of science courses.

Dave Ando
Dartford, UK

from aqueous samples. Particular emphasis is placed on liquid–liquid extraction, with reference to both ion-exchange and co-precipitation. The second part of this chapter is focused on the methods available for converting a solid sample into the appropriate form for elemental analysis. The most popular methods are based on acid digestion of the solid matrix, using either a microwave oven or a hot-plate approach. Details are provided on the methods available for the selective extraction of metal species in soil studies, using either single extraction or sequential extraction procedures. Finally, the role of enzymatic digestion procedures for food and soil matrices is described.

Chapter 3 explores the different approaches available for the introduction of samples into an inductively coupled plasma. While the most common approach uses the generic nebulizer/spray chamber arrangement, the choice of which nebulizer and/or spray chamber requires an understanding of the principle of operation and the benefits of each design. Alternative approaches for discrete sample introduction are also discussed, including a brief overview of electrothermal vaporization, followed by a detailed explanation of laser ablation and its issues concerning calibration. The flexibility of flow injection and/or chromatography for sample manipulation 'on-line' is highlighted, as well as the opportunities for introducing gaseous forms of metals/metalloids using hydride generation and/or cold vapour techniques.

Chapter 4 describes the principle of operation of an inductively coupled plasma. The concept of viewing position is also introduced. This is of particular importance in atomic emission spectroscopy where the plasma can be viewed either side-on or axially. For completeness, other plasma sources used in atomic spectroscopy are also discussed and highlighted, including the direct-current plasma, microwave-induced plasma and glow discharge.

Chapter 5 concentrates on the fundamental and practical aspects of inductively coupled plasma–atomic emission spectroscopy (ICP–AES). After an initial discussion of the fundamentals of spectroscopy as related to atomic emission spectroscopy, this chapter then focuses on the practical aspects of spectrometer design and detection. The ability to measure elemental information sequentially or simultaneously is discussed in terms of spectrometer design. Advances in detector technology, in terms of charge-transfer technology, are highlighted in the context of ICP–AES.

Chapter 6 describes the fundamental and practical aspects of inductively coupled plasma–mass spectrometry (ICP–MS). After an initial discussion of the fundamentals of mass spectrometry, this chapter then focuses on the types of mass spectrometer and variety of detectors available for ICP–MS. The occurrence of isobaric and molecular interferences in ICP–MS is highlighted, along with suggested remedies. Of particular note is a discussion of collision and reaction cells in ICP–MS. Emphasis is also placed on the capability of ICP–MS to perform quantitative analysis using isotope dilution analysis (IDA).

Preface

The technique of inductively coupled plasma (ICP) spectroscopy has expanded and diversified in the form of a mini-revolution in the last forty years. What was essentially an optical emission spectroscopic technique for trace element analysis has expanded into a source for both atomic emission spectroscopy and mass spectrometry, capable of detecting elements at sub-ppb (ng ml^{-1}) levels with good accuracy and precision. Modern instruments have shrunk in physical size, but expanded in terms of their analytical capabilities, so reflecting the significant developments in both optical and semiconductor technology.

The eight chapters contained in this book cover the following: the practical aspects of inductively coupled plasma spectroscopy, from the preparation of standards for instrument calibration, to the identification and selection of the most appropriate sample preparation techniques for solid and aqueous samples, the methods of sample introduction into an ICP, and the choice of atomic emission spectroscopy or mass spectrometry and their inherent advantages and disadvantages, finally offering guidance on the most effective approach for record keeping in the laboratory. The book is augmented with a chapter on selected applications describing the advantages that ICP technology can offer.

In Chapter 1, information is provided with regard to the general methodology for trace elemental analysis. This includes specific guidance on the recording of numerical data (with the appropriate units) with examples on how to display data effectively in the form of tables and figures. Issues relating to sample handling of solids and liquids are also covered. Numerical exercises involving the calculation of dilution factors and their use in determining original concentrations in aqueous and solid samples are provided as worked examples. Finally, the concept of quality assurance is introduced, together with the role of certified reference materials in trace element analysis.

Chapter 2 focuses on the specific sample preparation approaches for the elemental analysis of metals/metalloids from solid and aqueous samples. The first part of this chapter is concerned with methods for the extraction of metal ions

Confidence limit The extreme values or 'end-values' in a confidence interval.

Contamination In trace analysis this is the unintentional introduction of analyte(s) or other species which are not present in the original sample and which may cause an error in the determination. This can occur at any stage in the analysis. Quality assurance procedures, such as analyses of blanks or of reference materials, are used to check for contamination problems.

Control of Substances Hazardous to Health (COSHH) Regulations that impose specific legal requirements for risk assessment wherever hazardous chemicals or biological agents are used.

Co-precipitation The inclusion of otherwise soluble ions during the precipitation of lower-solubility species.

Dilution factor The mathematical factor applied to the determined value (data obtained from a calibration graph) which allows the concentration in the original sample to be determined. Frequently, for solid samples, this will involve a sample weight and a volume to which the digested/extracted sample is made up to prior to analysis. For liquid samples, this will involve an initial sample volume and a volume to which the digested/extracted sample is made up to prior to analysis.

Dissolved Material that will pass through a $0.45\,\mu\text{m}$ membrane filter assembly prior to sample acidification.

Dry ashing Use of heat to destroy the organic matrix of a sample to liberate the metal content.

Error The error of an analytical result is the difference between the result and a 'true' value:

Random error Result of a measurement minus the mean that would result from an infinite number of measurements of the same measurand carried out under repeatability conditions.

Systematic error Mean that would result from an infinite number of measurements of the same measurand carried out under repeatability conditions, minus the true value of the measurand.

Extraction The removal of a soluble material from a solid mixture by means of a solvent or the removal of one or more components from a liquid mixture by use of a solvent with which the liquid is immiscible or nearly so.

Figure of merit A parameter that describes the quality of performance of an instrument or an analytical procedure.

'Fitness for purpose' The degree to which data produced by a measurement process enables a user to make technically and administratively correct decisions for a stated purpose.

Heterogeneity The degree to which a property or a constituent is randomly distributed throughout a quantity of material. The degree of heterogeneity is the determining factor of sampling error.

be carried out. For trace element analysis, it is possible to purchase commercial systems that filter (distilled) water through a combination of ion-exchange columns to remove trace element impurities.

The cleaned vessels should then either be stored upside down or covered with Clingfilm® to prevent dust contamination.

The individual working in the laboratory is also a major source of contamination. Therefore, as well as the normal laboratory safety associated with wearing a laboratory coat and safety glasses, it may be necessary to take additional steps such as the wearing of 'contaminant-free' gloves and a close-fitting hat.

1.2 Analytical Terms and their Definitions

The following is an alphabetical list of the most important analytical terms of use in practical inductively coupled plasma spectroscopy.

Accuracy A quantity referring to the difference between the mean of a set of results or an individual result and the value which is accepted as the true or correct value for the quantity being measured.

Acid digestion Use of acid (and often heat) to destroy the organic matrix of a sample to liberate the metal content.

Aliquot A known amount of a homogenous material assumed to be taken with negligible sampling error.

Analyte The component of a sample which is ultimately determined directly or indirectly.

Bias Characterizes the systematic error in a given analytical procedure and is the (positive or negative) deviation of the mean analytical result from the (known or assumed) true value.

Calibration The set of operations which establish, under specified conditions, the relationship between values indicated by a measuring instrument or measuring system and the corresponding known values of the measurand.

Calibration curve Graphical representation of a measuring signal as a function of quantity of analyte.

Certified Reference Material (CRM) Reference material, accompanied by a certificate, one or more of whose property values are certified by a procedure which establishes its traceability to an accurate realization of the units in which the property values are expressed, and for which each certified value is accompanied by an uncertainty at a stated level of confidence.

Complexing agent The chemical species (an ion or a compound) which will bond to a metal ion using lone pairs of electrons.

Confidence interval Range of values that contains the true value at a given level of probability. The latter is known as the *confidence level*.

requires knowledge of a whole range of disciplines that need to come together to create the final result. The disciplines required can be described as follows:

- sampling, sample storage and preservation, and sample preparation methodologies
- analytical technique
- data control, including calibration strategies and the use of certified reference materials for quality control
- data management, including reporting of results and their meaning

While most of these are covered to some extent in this book, the reader should also consult other resources, e.g. books. Suggestions for further study are given in Chapter 8.

The perspective that is required when faced with trace element analysis are the additional precautions required in terms of management of contamination, choice of reagents and acids, and cleanliness of the workspace. For example, the grade of chemical used to prepare calibration standards is a major concern when working at trace element analysis levels (sub-μg ml^{-1}). Chemicals are available in a range of grades from 'GPR – general purpose reagent' through to, for example, 'AnalaR$^®$ – analytical reagent'. However, the descriptor does not identify that the chemical is any purer (the purity is often given on the bottle, e.g. 99.8%), but identifies the amount of effort that the manufacturer has gone to in the preparation of the chemical. For 'AnalaR$^®$' grade materials, the manufacturer has characterized the chemical by subjecting it to chemical analysis. This additional effort is then passed on to the customer in the form of a higher price. The use of sample blanks in the analytical procedure is essential to identify 'problem elements'.

The risk of contamination is a major problem in trace element analysis. Apart from the analytical reagent used to prepare standards, as discussed above, contamination can also be experienced from sample containers, e.g. volumetric flasks, pipettes, etc. For example, metal ions can adsorb onto glass containers and then leach into the solution under acidic conditions, thereby causing contamination. This can be minimized by cleaning the glassware prior to use by soaking for at least 24 h in a 10% nitric acid solution, followed by rinsing with 'clean' deionized water (three times).

DQ 1.1

What type of water would you consider to be clean?

Answer

Drinking water, commercially bottled drinking water or laboratory distilled water. The actual answer depends to some extent on the work to

Chapter 1

Methodology for Trace Elemental Analysis

Learning Objectives

- To be aware of the different types of contamination that can cause problems in trace elemental analysis.
- To be aware of a whole range of terms and definitions as used in analytical chemistry.
- To appreciate the range of units used in analytical chemistry.
- To be able to present numerical data with correct units, and be able to interchange the units as required.
- To be able to present data in the form of a table.
- To be able to present data in the form of a graph.
- To be able to determine the concentration of an element from a straight line graph using the equation $y = mx + c$.
- To be able to calculate the dilution factor for a liquid sample and a solid sample, and hence determine the concentration of the element in the original sample.
- To appreciate the concept of quality assurance in the analytical laboratory.
- To be aware of the significance of certified reference materials in elemental analysis.

1.1 Introduction

Trace elemental analysis requires more than just knowledge of the analytical technique to be used – in this case, inductively coupled plasma spectroscopy. This

Practical Inductively Coupled Plasma Spectroscopy   J. R. Dean
© 2005 John Wiley & Sons, Ltd

About the Author

John R. Dean, B.Sc., M.Sc., Ph.D., D.I.C., D.Sc., FRSC, CChem., CSci., Cert. Ed., Registered Analytical Chemist

John R. Dean took his first degree in Chemistry at the University of Manchester Institute of Science and Technology (UMIST), followed by an M.Sc. in Analytical Chemistry and Instrumentation at Loughborough University of Technology, and finally a Ph.D. and D.I.C. in Physical Chemistry at the Imperial College of Science and Technology (University of London). He then spent two years as a postdoctoral research fellow at the Food Science Laboratory of the Ministry of Agriculture, Fisheries and Food in Norwich, in conjunction with the Polytechnic of the South West in Plymouth. His work there was focused on the development of directly coupled high performance liquid chromatography and inductively coupled plasma–mass spectrometry methods for trace element speciation in foodstuffs. This was followed by a temporary lectureship in Inorganic Chemistry at Huddersfield Polytechnic. In 1988, he was appointed to a lectureship in Inorganic/Analytical Chemistry at Newcastle Polytechnic (now Northumbria University). This was followed by promotion to Senior Lecturer (1990), Reader (1994), Principal Lecturer (1998) and Associate Dean (Research) (2004). He was also awarded a personal chair in 2004. In 1998, he was awarded a D.Sc. (University of London) in Analytical and Environmental Science and was the recipient of the 23rd Society for Analytical Chemistry (SAC) Silver Medal in 1995. He has published extensively in analytical and environmental science. He is an active member of The Royal Society of Chemistry (RSC) Analytical Division, having served as a member of the atomic spectroscopy group for 15 years (10 as Honorary Secretary), as well as a past Chairman (1997–1999). He has served on the Analytical Division Council for three terms and is a former Vice-President (2002–2004), as well as a past-Chairman of the North-East Region of the RSC (2001–2003).

RSD	relative standard deviation
rms	root mean square
SAX	strong anion exchange
SCI	Society of Chemical Industry
SCX	strong cation exchange
SD	standard deviation
SE	standard error
SFMS	sector-field mass spectrometry
SI (units)	Système International (d'Unitès) (International System of Units)
TOF	time-of-flight
URL	uniform resource locator
USEPA	United States Environmental Protection Agency
USNLM	United States National Library of Medicine
UV	ultraviolet
V	volt
W	watt
WWW	World Wide Web

c	speed of light; concentration
e	electronic charge
E	energy; electric field strength
f	(linear) frequency; focal length
F	Faraday constant
h	Planck constant
I	intensity; electric current
m	mass
M	spectral order
p	pressure
Q	electric charge (quantity of electricity)
R	resolution; correlation coefficient; molar gas constant; resistance
R^2	coefficient of determination
t	time; Student factor
T	thermodynamic temperature
V	electric potential
z	ionic charge

λ	wavelength
ν	frequency (of radiation)
σ	measure of standard deviation
σ^2	variance

ESA	electrostatic analyzer
ETV	electrothermal vaporization
eV	electronvolt
FAAS	flame atomic absorption spectroscopy
GC	gas chromatography
GFAAS	graphite-furnace atomic absorption spectroscopy
HPLC	high performance liquid chromatography
Hz	hertz
IC	ion chromatography
ICP	inductively coupled plasma
ICP–AES	inductively coupled plasma–atomic emission spectroscopy
ICP–MS	inductively coupled plasma–mass spectrometry
id	internal diameter
IDA	isotope dilution analysis
IUPAC	International Union of Pure and Applied Chemistry
J	joule
KE	kinetic energy
LC	liquid chromatography
LDR	linear dynamic range
LGC	Laboratory of the Government Chemist
LLE	liquid–liquid extraction
LOD	limit of detection
LOQ	limit of quantitation
M_r	relative molecular mass
MAE	microwave-accelerated extraction
MDL	minimum detectable level
MIBK	methylisobutyl ketone
MIP	microwave-induced plasma
MS	mass spectrometry
MSD	mass-selective detector
N	newton
NIST	National Institute of Standards and Technology
Pa	pascal
PBET	physiologically based extraction test
PFE	pressurized fluid extraction
PGE(s)	platinum group element(s)
pMAE	pressurized microwave-accelerated extraction
PMT	photomultiplier tube
ppb	parts per billion (10^9)
ppm	parts per million (10^6)
ppt	parts per thousand (10^3)
RF	radiofrequency
RSC	The Royal Society of Chemistry

Acronyms, Abbreviations and Symbols

A_r	relative atomic mass
AAS	atomic absorption spectroscopy
AC	alternating current
ACS	American Chemical Society
A/D	analogue-to-digital
AES	atomic emission spectroscopy
aMAE	atmospheric microwave-accelerated extraction
ANOVA	analysis of variance
APDC	ammonium pyrrolidine dithiocarbamate
ASE™	accelerated solvent extraction
BEC	background equivalent concentration
BIDS	Bath Information and Data Services
C	coulomb
CCD	charge-coupled device; central composite design
CID	charge-injection device
COSSH	Control of Substances Hazardous to Health (Regulations)
CoV	coefficient of variation
CRM	Certified Reference Material
CTD	charge-transfer device
Da	dalton (atomic mass unit)
DAD	diode-array detection
DC	direct current
DCM	dichloromethane
DCP	direct-current plasma
DTPA	diethylenetriamine pentaacetic acid
EDTA	ethylenediamine tetraacetic acid

Acknowledgements

This present text includes material which has previously appeared in two of the author's earlier books, i.e. *Atomic Absorption and Plasma Spectroscopy* (ACOL Series, 1997) and *Methods for Environmental Trace Analysis* (AnTS Series, 2003), both published by John Wiley & Sons, Ltd, Chichester, UK.

The author is grateful to the copyright holders for granting permission to reproduce figures and tables from his two earlier publications.

Chapter 7 focuses on the applications of ICP technology. Examples have been selected from the scientific literature to highlight the importance and diversity of applications of ICP technology in analytical science. The particular areas highlighted are forensic science (document analysis), industrial analysis (coal), materials analysis (gadolinium oxide), environmental analysis (soil), food analysis (milk products) and pharmaceutical analysis.

The final chapter (Chapter 8) provides examples of data sheets that can be used to record information in the laboratory. Guidelines are given for the recording of information associated with sample pre-treatment. Then, specific sheets are provided for the recording of sample preparation details, as well as the instrumental aspects of ICP–AES and ICP–MS. This chapter concludes with guidance on the range of resources available in paper and electronic format to assist in the understanding of ICP technology and its application to trace element analysis.

John R. Dean
Northumbria University, Newcastle, UK

Homogeneity The degree to which a property or a constituent is uniformly distributed throughout a quantity of material. A material may be homogenous with respect to one analyte but heterogeneous with respect to another.

Interferent Any component of the sample affecting the final measurement.

Limit of detection The detection limit of an individual analytical procedure is the lowest amount of an analyte in a sample which can be detected but not necessarily quantified as an exact value. The limit of detection, expressed as either the concentration c_L or the quantity q_L, is derived from the smallest measure, x_L, that can be detected with reasonable certainty for a given procedure. The value x_L is given by the following equation:

$$x_L = x_{bl} + k s_{bl}$$

where x_{bl} is the mean of the blank measures, s_{bl} is the standard deviation of the blank measures and k is a numerical factor chosen according to the confidence level required. For many purposes, the limit of detection is taken to be $3 s_{bl}$ or $3 \times$ 'the signal-to-noise ratio', assuming a zero blank.

Limit of quantitation For an individual analytical procedure, this is the lowest amount of an analyte in a sample which can be quantitatively determined with suitable uncertainty. It may also be referred to as the *limit of determination*. The limit of quantitation can be taken as $10 \times$ 'the signal-to-noise ratio', assuming a zero blank.

Linear dynamic range (LDR) The concentration range over which the analytical working calibration curve remains linear.

Linearity This defines the ability of the method to obtain test results proportional to the concentration of analyte.

Liquid–liquid extraction A method of extracting a desired component from a liquid mixture by bringing the solution into contact with a second liquid, the solvent, in which the component is also soluble, and which is immiscible with the first liquid or nearly so.

Matrix The carrier of the test component (analyte), all of the constituents of the material except the analyte, or the material with as low a concentration of the analyte as it is possible to obtain.

Measurand A particular quantity subject to measurement.

Method The overall, systematic procedure required to undertaken an analysis. This includes all stages of the analysis, and not just the (instrumental) end determination.

Microwave digestion A method of digesting an organic matrix to liberate metal content by using an acid at elevated temperature (and pressure) based on microwave radiation. Can be carried out in either open or sealed vessels.

Organometallic An organic compound in which a metal is covalently bonded to carbon.

Outlier This may be defined as an observation in a set of data that appears to be inconsistent with the remainder of that set.

Precision The closeness of agreement between independent test results obtained under stipulated conditions.

Qualitative analysis Chemical analysis designed to identify the components of a substance or mixture.

Quality assurance All those planned and systematic actions necessary to provide adequate confidence that a product or services will satisfy given requirements for quality.

Quality control The operational techniques and activities that are used to fulfill requirements of quality.

Quality control chart A graphical record of the monitoring of control samples which helps to determine the reliability of the results.

Quantitative analysis This is normally taken to mean the numerical measurement of one or more analytes to the required level of confidence.

Reagent A test substance that is added to a system in order to bring about a reaction or to see whether a reaction occurs (e.g. an analytical reagent).

Reagent blank A solution obtained by carrying out all steps of the analytical procedure in the absence of a sample.

Recovery The fraction of the total quantity of a substance recoverable following a chemical procedure.

Reference material This is a material or substance, one or more of whose property values are sufficiently homogeneous and well established to be used for the calibration of an apparatus, the assessment of a measurement method, or for assigning values to materials.

Repeatability Precision under repeatability conditions, i.e. conditions where independent test results are obtained with the same method on identical test items in the same laboratory, by the same operator, using the same equipment within short intervals of time.

Reproducibility Precision under reproducibility conditions, i.e. conditions where test results are obtained with the same method on identical test items in different laboratories, with different operators, using different equipment.

Robustness For an analytical procedure, this is a measure of its capacity to remain unaffected by small, but deliberate variations in method parameters, and provides an indication of its reliability during normal usage. It is sometimes referred to as *ruggedness*.

Sample A portion of material selected from a larger quantity of material. The term needs to be qualified, e.g. representative sample, sub-sample, etc.

Selectivity (in analysis) Qualitative – the extent to which other substances interfere with the determination of a substance according to a given procedure. Quantitative – a term used in conjunction with another substantive (e.g. constant, coefficient, index, factor, number, etc.) for the quantitative characterization of interferences.

Signal-to-noise ratio A measure of the relative influence of noise on a control signal. Usually taken as the magnitude of the signal divided by the standard deviation of the background signal.

Solvent extraction The removal of a soluble material from a solid mixture by means of a solvent or the removal of one or more components from a liquid mixture by use of a solvent with which the liquid is immiscible or nearly so.

Speciation The process of identifying and quantifying the different defined species, forms or phases present in a material or the description of the amounts and types of these species, forms or phases present.

Standard (all types) A standard is an entity established by consensus and approved by a recognized body. It may refer to a material or solution (e.g. an organic compound of known purity or an aqueous solution of a metal of agreed concentration) or a document (e.g. a methodology for an analysis or a quality system). The relevant terms are as follows:

Analytical standard (*also known as* **Standard solution**) A solution or matrix containing the analyte which will be used to check the performance of the method/instrument.

Calibration standard The solution or matrix containing the analyte (measurand) at a known value with which to establish a corresponding response from the method/instrument.

External standard A measurand, usually identical with the analyte, analysed separately from the sample.

Internal standard A measurand, similar to but not identical with the analyte, which is combined with the sample.

Standard method A procedure for carrying out a chemical analysis which has been documented and approved by a recognized body.

Standard addition The addition of a known amount of analyte to the sample in order to determine the relative response of the detector to an analyte within the sample matrix. The relative response is then used to assess the sample analyte concentration.

Stock solution This is generally a standard or reagent solution of known accepted stability, which has been prepared in relatively large amounts of which portions are used as required. Frequently, such portions are used following further dilution.

Sub-sample This may be either (a) a portion of the sample obtained by selection or division, (b) an individual unit of the lot taken as part of the sample, or (c) the final unit of multi-stage sampling.

True value A value consistent with the definition of a given particular quantity.

Uncertainty A parameter associated with the result of a measurement which characterizes the dispersion of the values that could reasonably be attributed to the measurand.

1.3 Units

The Systeme International d'Unites (SI) uses a series of base units (Table 1.1) from which other terms have been derived. Some of the most commonly used SI derived units are shown in Table 1.2. When using units, it is standard practice to keep numbers between 0.1 and 1000 by the use of a set of prefixes, based on multiples of 10^3 (Table 1.3). It is an extremely useful skill to be able to interchange these units and prefixes. For example, $1 \, mol \, l^{-1}$ can also be expressed as $1000 \, \mu mol \, ml^{-1}$, $1000 \, mmol \, l^{-1}$ or $1000 \, nmol \, \mu l^{-1}$.

SAQ 1.1

The prefixes shown below in Table 1.3 are frequently used in analytical science to represent large or small quantities. Reassign the following quantities with the suggested prefixes.

Quantity	m	μm	nm
$3 \times 10^{-7} \, m$			
Quantity	$mol \, l^{-1}$	$mmol \, l^{-1}$	$\mu mol \, l^{-1}$
$6.9 \times 10^{-3} \, mol \, l^{-1}$			
Quantity	$\mu g \, ml^{-1}$	$mg \, l^{-1}$	$ng \, \mu l^{-1}$
$2.80 \, ppm$			

1.4 Calibration Strategies

Quantitative analysis in plasma spectroscopy requires the preparation of a series of calibration standards from a stock solution. These standards are prepared in

Table 1.1 Commonly used base SI Units. From Dean, J. R., *Methods for Environmental Trace Analysis*, AnTS Series. Copyright 2003. © John Wiley and Sons, Limited. Reproduced with permission

Measured quantity	Name of unit	Symbol
Length	Metre	m
Mass	Kilogram	kg
Amount of substance	Mole	mol
Time	Second	s
Thermodynamic temperature	Kelvin	K
Luminous intensity	Candela	cd

Table 1.2 Commonly used SI derived units. From Dean, J. R., *Methods for Environmental Trace Analysis*, AnTS Series. Copyright 2003. © John Wiley and Sons, Limited. Reproduced with permission

Measured quantity	Name of unit	Symbol	Definition in base units	Alternative in derived units
Energy	Joule	J	m^2 kg s^{-2}	N m
Force	Newton	N	m kg s^{-2}	J m^{-1}
Pressure	Pascal	Pa	kg m^{-1} s^{-2}	N m^{-2}
Electric charge	Coulomb	C	A s	J V^{-1}
Frequency	Hertz	Hz	s^{-1}	—

Table 1.3 Commonly used prefixes. From Dean, J. R., *Methods for Environmental Trace Analysis*, AnTS Series. Copyright 2003. © John Wiley and Sons, Limited. Reproduced with permission

Multiple	Prefix	Symbol
10^{18}	exa	E
10^{15}	peta	P
10^{12}	tera	T
10^9	giga	G
10^6	mega	M
10^3	kilo	k
10^{-3}	milli	m
10^{-6}	micro	μ
10^{-9}	nano	n
10^{-12}	pico	p
10^{-15}	femto	f
10^{-18}	atto	a

volumetric flasks. Calibration solutions are usually prepared in terms of their molar concentrations, i.e. mol l^{-1}, or mass concentrations, i.e. g l^{-1}, with both referring to an amount per unit volume, i.e. concentration = amount/volume. It is important to use the highest (purity) grade of chemicals (liquids or solids) for the preparation of the stock solution, e.g. AnalaR® or AristaR®.

DQ 1.2

How would you prepare a 1000 ppm solution of lead from lead nitrate?

Answer

The molecular weight of lead nitrate, $Pb(NO_3)_2$, is 331.20 and the atomic weight of lead is 207.19. Therefore, simply dividing the molecular weight *by the atomic weight will give you the exact amount of lead nitrate to be dissolved in 1 litre to produce a 1000 ppm solution of lead.*

$$331.20/207.19 = 1.5985 \ g \ of \ Pb(NO_3)_2 \ in \ 1 \ litre$$

Frequently, however, a litre of stock solution is not required.

DQ 1.3

What mass of lead nitrate would need to be accurately weighed to pre-pare 100 ml of the 1000 ppm stock solution?

Answer

With reference to DQ 1.2, the mass is simply obtained by dividing the amount in 1 litre by 10, i.e. 0.1599 g.

1.5 Presentation of Data: Tables

It is often convenient when carrying out quantitative laboratory work to record data in tabular form. This is often best done by creating two columns into which the data can be entered. It is essential, however, for future consultation of these data, that the columns are given the appropriate headings, e.g. concentration ($\mu g \, l^{-1}$) and signal (mV), to prevent errors occurring later. It is also important to record details of any sample dilutions that have taken place (see also Chapter 8). A typical table of data for an experiment to determine the concentration of lead is shown in Table 1.4. It is perfectly acceptable to record information, by using a pen, in a laboratory notebook. Any modifications should then be carried out by simply crossing out the erroneous data and entering the correct information.

Table 1.4 An example of how to record quantitative data for an inductively coupled plasma–mass spectrometry (ICP–MS) experiment

Concentration ($\mu g \, l^{-1}$)	^{208}Pb intensity (counts s^{-1})
0	565
10	19 887
20	45 356
30	59 876
40	78 543
50	99 654

1.6 Presentation of Data: Graphs

Most graphs are now plotted by using computer-based graphics packages, e.g. Microsoft Excel™, rather than by hand on graph paper. You should remember, however, that you still need the skill to plot a graph by hand on graph paper as this is still the most common method used in examinations. Irrespective of the mode of preparing the graph, it is important to ensure that the graph is correctly labelled and presented. All graphs should have a numerical descriptor and title, e.g. 'Figure 3.2 Influence of time (hours) on the recovery of metal species from a sediment sample'.

Graphs are normally used to describe a relationship between two variables, e.g. x and y. It is normal practice to identify the x-axis as the horizontal axis (absicca) and to use this for the independent variable, e.g. concentration (with its appropriate units). The vertical axis or ordinate (y-axis) is used to plot the dependent variable, e.g. signal response (with units, if appropriate). The mathematical relationship most commonly used for straight-line graphs is as follows:

$$y = mx + c$$

where y is the signal response, e.g. signal (mV), x is the concentration of the working solution (in appropriate units, e.g. $\mu g\ ml^{-1}$ or ppm), m is the slope of the graph, and c is the intercept on the x-axis.

A typical graphical representation of data obtained from an experiment to determine the level of lead in a sample using inductively coupled plasma–mass spectrometry (ICP–MS) is shown in Figure 1.1 (from the data tabulated in Table 1.4).†

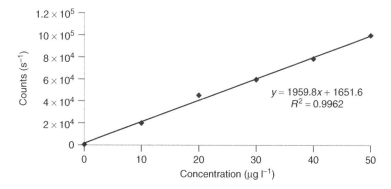

Figure 1.1 Typical calibration graph used to determine lead (^{208}Pb) in a sample (cf. Table 1.4).

---

† R, in this figure (and also in Figure 1.2), is known as the *correlation coefficient*, and provides a measure of the quality of calibration. In practice, R^2 (the *coefficient of determination*) is used because it is more sensitive to changes. This varies between -1 and $+1$, with values very close to -1 and $+1$ pointing to a very tight 'fit' of the calibration curve.

An alternative approach to undergoing a *direct calibration*, as described above, is the use of the method of *standard additions*. This may be particularly useful if the sample is known to contain a significant portion of a potentially interfering matrix. In standard additions, the calibration plot no longer passes through zero (on both the x- and y-axes). As the concept of standard additions is to eliminate any matrix effects present in the sample, it should be implicit that the working standard solutions all will contain the same volume of the sample, i.e. the same volume of the sample solution is introduced into a succession of working calibration solutions. Each of these solutions, containing the same volume of the sample, is then introduced into the inductively coupled plasma and the response recorded. However, plotting the signal response (e.g. signal (mV)) against analyte concentration produces a graph that no longer passes through zero on either axis, but if correctly drawn, the graph can be extended towards the x-axis (extrapolated) until it intercepts it. By maintaining a constant concentration x-axis, the unknown sample concentration can be determined (Figure 1.2). It is essential that this graph is linear over its entire length or otherwise considerable error can be introduced.

In both the direct calibration graph and the method of additions graph, the result obtained from the sample is normally not the final answer of how much of the metal was in the original sample. This is because the sample has normally undergone some form of sample preparation. In the case of a solid sample, this might have involved acid digestion (see Section 2.3.1), while in the case of a liquid sample, liquid–liquid extraction (see Section 2.2.1). What is required in both cases is a correction (or dilution factor) that takes into account the sample

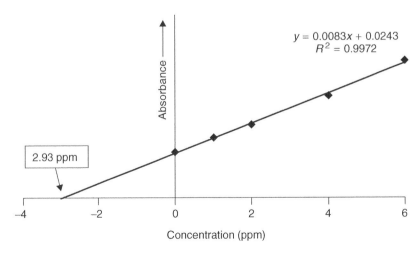

Figure 1.2 Determination of lead in soil: standard additions method. From Dean, J. R., *Methods for Environmental Trace Analysis*, AnTS Series. Copyright 2003. © John Wiley & Sons, Limited. Reproduced with permission.

preparation procedure. The following provides examples of the general forms of calculations that are necessary in the case of (1) a liquid sample that has been extracted using ammonium pyrrolidine dithiocarbamate (APDC)–methylisobutyl ketone (MIBK), and (2) a solid sample that has been acid-digested (see Figures 2.1 and 2.3, respectively, for further details of the procedures).

1.7 Calculations: Dilution Factors

Case study 1. Calculate the concentration ($\mu g\, ml^{-1}$) of copper in a waste water sample obtained from the local waste treatment plant. A waste water sample (150 ml) was extracted with APDC–diethylammonium diethyldithiocarbamate (DDDC) into MIBK (20 ml) using liquid–liquid extraction. The extract was then quantitatively transferred to a 25 ml volumetric flask and made up to the mark with MIBK. What is the dilution factor?

$$25\, ml/150\, ml = 0.167\, ml\, ml^{-1}$$

If the solution was then analysed and found to be within the linear portion of the graph (see Figure 1.1), the value for the dilution factor should then be multiplied by the concentration from the graph, hence producing a final value indicating the concentration of copper in the waste water sample.

Case study 2. Calculate the concentration ($\mu g\, g^{-1}$) of lead in a soil sample obtained from a contaminated land site. An accurately weighed (5.2456 g) soil sample is acid-digested (see Figure 2.3) using nitric acid and hydrogen peroxide, cooled and then quantitatively transferred to a 100 ml volumetric flask and made up to the mark with distilled water. This solution is then diluted by taking 10 ml and transferring to a further 100 ml volumetric flask where it is made up to the mark with high-purity water. What is the dilution factor?

$$[(100\, ml)/(5.2456\, g) \times (100\, ml/10\, ml)] = 190.64\, ml\, g^{-1}$$

If the solution was then analysed and found to be within the linear portion of the graph (see Figure 1.1), the value for the dilution factor should then be multiplied by the concentration from the graph, so producing a final value indicating the concentration of lead in the contaminated soil sample.

1.8 Quality Assurance and the Use of Certified Reference Materials

Quality assurance is all about getting the correct result. The main objectives of a quality assurance scheme are as follows:

• to select and validate appropriate methods of sample preparation

National Institute of Science and Technology
Certificate of Analysis

Standard Reference Material 1515
Apple Leaves

Certified Concentrations of Constituent Elements[1]

Element	Concentration (wt%)
Calcium	1.526 ± 0.015
Magnesium	0.271 ± 0.008
Nitrogen (total)	2.25 ± 0.19
Phosphorus	0.159 ± 0.011
Potassium	1.61 ± 0.02

Element	Concentration $(\mu g\ g^{-1})$[2]	Element	Concentration $(\mu g\ g^{-1})$[2]
Aluminium	286 ± 9	Mercury	0.044 ± 0.004
Arsenic	0.038 ± 0.007	Molybdenum	0.094 ± 0.013
Barium	49 ± 2	Nickel	0.91 ± 0.12
Boron	27 ± 2	Rubidium	10.2 ± 1.5
Cadmium	0.013 ± 0.002	Selenium	0.050 ± 0.009
Chlorine	579 ± 23	Sodium	24.4 ± 12
Copper	5.64 ± 0.24	Strontium	25 ± 2
Iron	83 ± 5	Vanadium	0.26 ± 0.03
Lead	0.470 ± 0.024	Zinc	12.5 ± 0.3
Manganese	54 ± 3		

[1]The certified concentrations are equally weighted means of results from two or more different analytical methods or the means of results from a single method of known high accuracy.

[2]The values are based on dry weights. Samples of this SRM must be dried before weighing and analysis by, for example, drying in a desiccator at room temperature (ca. 22 °C) for 120 h over fresh anhydrous magnesium perchlorate. The sample depth should not exceed 1 cm.

Figure 1.3 An example of a certificate of analysis for elements in apple leaves. Reprinted from Certificate of Analysis, *Standard Reference Material 1515, Apple Leaves*, National Institute of Standards and Technology. Not copyrightable in the United States.

- to select and validate appropriate methods of analysis

- to maintain and upgrade analytical instruments

- to ensure good record-keeping of methods and results

- to ensure quality of the data produced
- to maintain a high quality of laboratory performance

In implementing a good quality control programme it is necessary to analyse *certified reference materials*. By definition, a certified reference material is a substance for which one or more elements have known values and estimates of their uncertainties, produced by a technically valid procedure, accompanied with a traceable certificate and issued by a certifying body. Typical examples of certifying bodies are the National Institute for Standards and Technology (NIST), based in Washington, DC, USA, the Community Bureau of Reference (BCR), Brussels, Belgium, and the Laboratory of the Government Chemist (LGC), London, UK. The accompanying certificate, in addition to providing details of the certified elemental concentration and their uncertainties in the sample, also provides details of the minimum sample weights to be used, storage conditions, moisture content, etc. An example of a typical certificate is shown in Figure 1.3.

Summary

The methodology for trace elemental analysis requires an understanding of a whole range of inter-related issues centred around the sample, sample preparation, analysis, data interpretation/presentation and quality assurance. This chapter has highlighted some of the most important aspects. In addition, the main strategies for calibration are discussed, including the preparation of a standard solution. Two examples of how to calculate a dilution factor are given for aqueous and solid materials.

Chapter 2

Sample Preparation for Inductively Coupled Plasma Spectroscopy

Learning Objectives

- To appreciate the different approaches available for preparation of samples for inductively coupled plasma spectroscopy.
- To appreciate the range of approaches available for sample pre-treatment of aqueous samples.
- To be aware of the most important variables in liquid–liquid extraction.
- To understand the concept of pre-concentration using ion-exchange chromatography.
- To be aware of the term 'co-precipitation' and its role in the pre-concentration of metals in aqueous samples.
- To be aware of the procedures for carrying out acid-digestion (hot-plate and microwave) in a safe and controlled manner.
- To be aware of other decomposition methods, e.g. fusion and dry ashing.
- To understand the relevance of selective extraction methods for soil studies.
- To be able to carry out single extraction procedures using ethylenediamine tetraacetic acid (EDTA), acetic acid and diethylenetriamine pentaacetic acid (DTPA) in a safe and controlled manner.
- To be able to carry out a sequential extraction procedure on a soil sample in a safe and controlled manner.
- To understand the potential for chemical speciation using enzyme digestion.

Practical Inductively Coupled Plasma Spectroscopy   J. R. Dean
© 2005 John Wiley & Sons, Ltd

2.1 Introduction

The ideal sample preparation for ICP analysis is none at all! However, even for aqueous samples this is not always the case. Often for aqueous samples, some form of filtration or pre-concentration is required. For solid samples, the normal practice is to destroy the sample matrix and thereby liberate the metals present. This chapter investigates the different approaches used for aqueous and solid samples with particular emphasis on the requirements for ICP analysis.

2.2 Aqueous Samples

The simplest form of sample preparation for aqueous samples, e.g. natural waters, is to filter the sample through a $0.2\,\mu m$ membrane to remove particulates prior to introduction into the plasma. However, even the most sensitive of analytical techniques, e.g. ICP–mass spectrometry (MS), may require some sample pretreatment, e.g. separation and/or pre-concentration. Remember that any form of separation and/or pre-concentration can also have the same effect on potential contaminants as well as the metal of interest. It is essential, therefore, to use (ultra) pure reagents and acids for treatment, as well as to take care in the cleaning of containers, e.g. beakers, volumetric flasks, etc., prior to use. The most important pre-concentration methods for metal ions are liquid–liquid extraction and ion-exchange.

2.2.1 Liquid–Liquid Extraction

The basis of liquid–liquid extraction (LLE) is to take a large volume of the aqueous solution containing the sample, e.g. natural water, form a metal complex by the addition of a suitable chelating agent, and then partition this complex into a small volume of an immiscible organic solvent – thereby allowing effective and quantitative pre-concentration of the metals present in the original sample into the organic solvent. The most influential variables in LLE are as follows:

- choice of chelating agent

- choice of organic solvent

- pH of the aqueous sample

The most commonly used chelating agent in atomic spectroscopy is ammonium pyrrolidine dithiocarbamate (APDC). This reagent, which is normally used as a 1–2% aqueous solution, can be directly extracted with an organic solvent, e.g. methylisobutyl ketone (MIBK). A description of a typical procedure to extract metals by using this approach is shown in Figure 2.1. As APDC is capable of

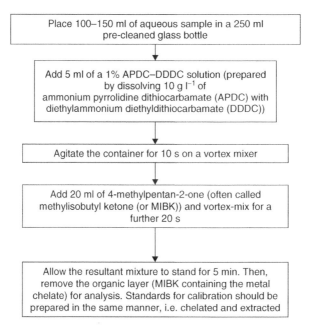

Figure 2.1 Typical procedure for the liquid–liquid extraction of metals [1]. From Dean, J. R., *Methods for Environmental Trace Analysis*, AnTS Series. Copyright 2003. © John Wiley & Sons, Limited. Reproduced with permission.

Table 2.1 Dependence of pH on the ability of ammonium pyrrolidine dithio-carbamate (APDC) to form complexes with various metals [2]. From Dean, J. R., *Methods for Environmental Trace Analysis*, AnTS Series. Copyright 2003. © John Wiley & Sons, Limited. Reproduced with permission

pH range	Metal–APDC complexes
2	W
2–4	Nb, U
2–6	As, Cr, Mo, V, Te
2–8	Sn
2–9	Sb, Se
2–14	Ag, Au, Bi, Cd, Co, Cu, Fe, Hg, Ir, Mn, Ni, Os, Pb, Pd, Pt, Ru, Rh, Tl, Zn

forming extractable metal complexes over a wide range of pH (Table 2.1), careful control allows some degree of selectivity to be achieved.

Some examples of other common chelating agents are shown in Figure 2.2. Of these, 8-hydroxyquinoline is particularly useful for extracting Al, Mg, Sr, V and W, diethyldithiocarbamates (e.g. the sodium derivative) for As(III), Bi, Sb(III),

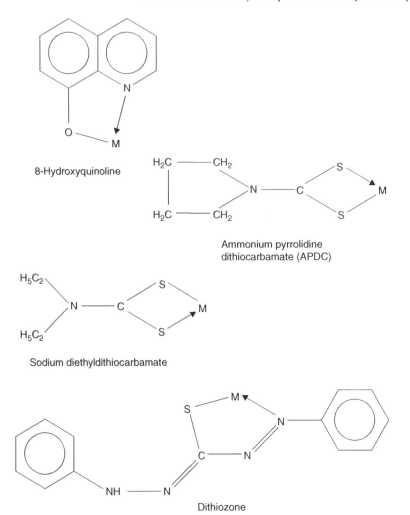

Figure 2.2 Some common metal chelates used for solvent extraction. Note that in all cases, the hydrogen ion of the parent chelating agent has been replaced by a metal [1]. From Dean, J. R., *Methods for Environmental Trace Analysis*, AnTS Series. Copyright 2003. © John Wiley & Sons, Limited. Reproduced with permission.

Se(IV), Sn(IV), Te(IV), Tl(III) and V(V), and dithiozone for Ag, Bi, Cu, Hg, Pb, Pd, Pt and Zn.

2.2.2 Ion-Exchange

Two types of ion-exchange media can be used, namely cation- or anion-exchange. As their names suggest, cation-exchange is used to separate metal ions (positively charged species), while anion-exchange is used to separate negatively charged species. Ion-exchange media can be used in two forms, i.e. batch mode or packed in columns. In batch mode, the ion-exchange medium is placed directly in the aqueous sample solution. After concentrating the metal ions, the ion-exchange media is then filtered to remove it from the aqueous sample prior to desorption. In column mode, the ion-exchange medium is packed into columns and the aqueous sample pumped through. The most popular mode in plasma analysis is column chromatography as this allows it to be used 'on-line' with direct coupling to the sample introduction system of the ICP (see also Chapter 3). At first hand, it may seem that the only useful form of ion-exchange medium is cation–exchange, but this is not always the case.

DQ 2.1

Can you suggest a situation where an anion-exchange medium may be used in metal determination?

Answer

In the case of arsenic determination, where arsenite, $AsO_2{}^-$, and arsenate, $AsO_4{}^{3-}$, can be pre-concentrated by using an anion-exchange medium.

In the general case of a strong cation-exchange resin, the following two stages can be identified, i.e. pre-concentration and desorption of metal ions.

$$n\mathrm{RSO_3}^-\mathrm{H}^+ + \mathbf{M}^{n+} = (\mathbf{RSO_3}^-)n\,\mathbf{M}^{n+} + \mathrm{H}^+ \tag{2.1}$$

$$(\mathbf{RSO_3}^-)n\,\mathbf{M}^{n+} + \mathrm{H}^+ = n\mathrm{RSO_3}^-\mathrm{H}^+ + \mathbf{M}^{n+} \tag{2.2}$$

In Equation (2.1), the metal ion (M^{n+}) in a large volume of aqueous solution is retained, i.e. pre-concentrated, on the cation-exchange resin ($n\mathrm{RSO_3}^-\mathrm{H}^+$), with the release of H^+. After passage of all of the aqueous water sample, the discrete addition of a small quantity of acid will release the concentrated metal ion from the cation-exchange resin $(\mathrm{RSO_3}^-)n\mathrm{M}^{n+}$, thereby regenerating the cation-exchange medium for the next sample (Equation (2.2)).

An alternative medium that allows the separation of an excess of alkali metal ions from other cations uses a chelating ion-exchange resin. This type of resin forms chelates or complexes with the metal ions. The most common of these is 'Chelex-100'. This resin contains iminodiacetic acid functional groups which act to 'trap' or complex metal ions. It has been found that Chelex-100, in acetate buffer at pH 5–6, can retain Al, Bi, Cd, Co, Cu, Fe, Ni, Pb, Mn, Mo, Sc, Sn, Th, U, V, W, Zn and Y, plus various rare-earth metals, while at the same time it does not retain alkali metals (e.g. Li, Na, Rb and Cs), alkali-earth metals (Be, Ca, Mg, Sr and Ba) and anions (F^-, Cl^-, Br^- and I^-).

2.2.3 Co-Precipitation

Co-precipitation allows the quantitative precipitation of the metal ion of interest by the addition of a co-precipitant. Co-precipitation of metal ions on collectors can be attributed to several mechanisms, as follows:

- *adsorption* – the charge on the surface can attract ions in solution of opposite charge
- *occlusion* – ions are embedded within the forming precipitate
- *co-crystallisation* – the metal ion can become incorporated in the crystal structure of the precipitate

The major disadvantage of co-precipitation is that the precipitate, which is present at a high mass-to-metal ratio, can be a major source of contamination. In addition, further sample preparation, e.g. dissolution, is required prior to analysis of the metal. This can also increase the risk of contamination and analyte losses. One of the most common co-precipitants is iron, e.g. as $Fe(OH)_3$.

2.3 Solid Samples

2.3.1 Decomposition Techniques

Acid-digestion involves the use of mineral or oxidizing acids and an external heat source to decompose the sample matrix. The choice of an individual acid or combination of acids is dependent upon the nature of the matrix to be decomposed (Table 2.2). The most obvious example of this relates to the digestion of a matrix containing silica (SiO_2), e.g. a geological material. In this situation, the only appropriate acid to digest the silica is hydrofluoric acid (HF). No other acid or combination of acids will liberate the metal of interest from the silica matrix.

SAQ 2.1

Why should hydrofluoric acid be so effective for the digestion of silica?

Once the choice of an acid is made, the sample is placed in an appropriate vessel for the decomposition stage. The choice of vessel, however, depends upon the nature of the heat source to be applied. Most commonly, the acid-digestion of solid matrices is carried out in open glass vessels (beakers or boiling tubes) by using a hot-plate or multiple-sample digestor. A typical procedure for the digestion of sediment, sludges and soil samples using a hot-plate is described in Figure 2.3. A multiple-sample digestor allows a number of boiling tubes (6, 12 or 24 tubes) to be placed into the well of a commercial digestor

Table 2.2 Some examples of common acids used for wet decomposition[a] [1]. From Dean, J. R., *Methods for Environmental Trace Analysis*, AnTS Series. Copyright 2003. © John Wiley & Sons, Limited. Reproduced with permission

Acid (s)	Boiling point (°C)	Comments
Hydrochloric (HCl)	110	Useful for salts of carbonates, phosphates, some oxides and some sulfides. A weak reducing agent; not generally used to dissolve organic matter
Sulfuric (H_2SO_4)	338	Useful for releasing a volatile product; good oxidizing properties for ores, metals, alloys, oxides and hydroxides; often used in combination with HNO_3. **Caution**: H_2SO_4 must never be used in PTFE vessels (PTFE has a melting point of 327°C and deforms at 260°C)
Nitric (HNO_3)	122	Oxidizing attack on many samples not dissolved by HCl; liberates trace elements as the soluble nitrate salt. Useful for dissolution of metals, alloys and biological samples
Perchloric ($HClO_4$)	203	At fuming temperatures, a strong oxidizing agent for organic matter. **Caution**: violent, explosive reactions may occur – care is needed. Samples are normally pre-treated with HNO_3 prior to addition of $HClO_4$
Hydrofluoric (HF)	112	For digestion of silica-based materials; forms SiF_6^{2-} in acid solution; **caution** is required in its use; glass containers should not be used, only plastic vessels. In case of spillages, calcium gluconate gel (for treatment of skin contact sites) should be available prior to useage; evacuate to hospital immediately if skin is exposed to liquid HF
Aqua regia (nitric/hydrochloric)	—	A 1:3 vol/vol mixture of HNO_3:HCl is called aqua regia; forms a reactive intermediate, NOCl. Used for metals, alloys, sulfides and other ores – best known because of its ability to dissolve Au, Pd and Pt

[a]Protective clothing/eyewear is essential in the use of concentrated acids. All acids should be handled with care and in a fumecupboard.

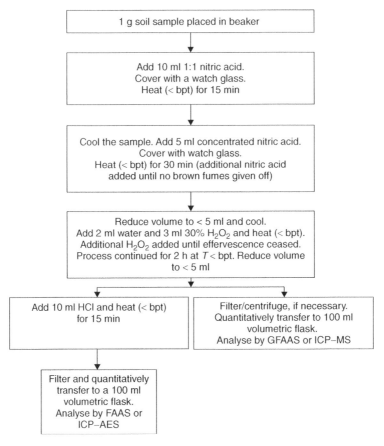

Figure 2.3 The United States Environmental Protection Agency (USEPA) procedure for the acid-digestion of sediments, sludges and soils using a hot-plate: GFAAS, graphite-furnace atomic absorption spectroscopy; FAAS, flame atomic absorption spectroscopy; ICP–MS, inductively coupled plasma–mass spectrometry; ICP–AES, inductively coupled plasma–atomic emission spectroscopy [1, 3]. From Dean, J. R., *Methods for Environmental Trace Analysis*, AnTS Series. Copyright 2003. © John Wiley & Sons, Limited. Reproduced with permission.

(Figure 2.4). An alternative approach to conventional heating involves the use of microwave heating.

2.3.2 Microwave Digestion

The first reported use of a microwave oven for the acid digestion of samples for metal analysis was in 1975 [4]. Advances in technology by a variety of manufacturers means that today that there are two types of microwave heating systems

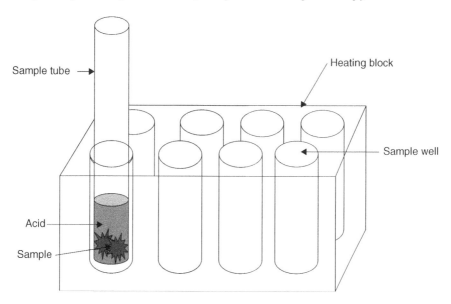

Figure 2.4 Schematic of a commercial acid-digestion system [1]. From Dean, J. R., *Methods for Environmental Trace Analysis*, AnTS Series. Copyright 2003. © John Wiley & Sons, Limited. Reproduced with permission.

commercially available, i.e. an open-focused and a closed-vessel system. In the open-style system, up to six sample vessels are heated simultaneously. A typical commercial system is the 'Simultaneous Temperature Accelerated Reaction' (STAR™) system from the CEM Corporation, USA. A schematic diagram of an open-focused microwave system is shown in Figure 2.5. The sample and acid (sulfuric acid can be used) are introduced into a glass container, which has the appearance of a large boiling/test tube, and is then fitted with a condensor to prevent loss of volatiles. The sample container is placed within the microwave cavity and heated. The use of open vessels for digestion can lead to additional problems associated with loss by volatilization of element species. This can be rectified by the correct choice of reagents used and the type of digestion apparatus employed.

A common commercial closed system is the 'Microwave Accelerated Reaction System' (MARS™), as supplied by the CEM Corporation, USA (Figure 2.6). This system allows up to 14 extraction vessels (XP-1500 Plus™) to be irradiated simultaneously. In addition, other features include a function for monitoring both pressure and temperature, and most notably, it is equipped with an alarm to call attention to an unexpected release of flammable and toxic material. The microwave energy output of this system is 1500 W at a frequency of 2450 MHz at 100% power. The pressure (up to 800 psi) is continuously measured (measurements taken at the rate of $200\,s^{-1}$), while the temperature (up to 300°C) is monitored for all vessels every 7 s. All of the sample vessels are held in a carousel

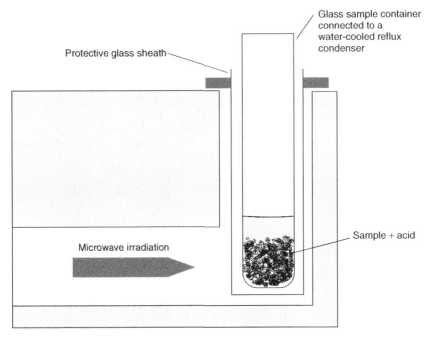

Figure 2.5 Schematic of an atmospheric open-focused microwave digestion system [1]. From Dean, J. R., *Methods for Environmental Trace Analysis*, AnTS Series. Copyright 2003. © John Wiley & Sons, Limited. Reproduced with permission.

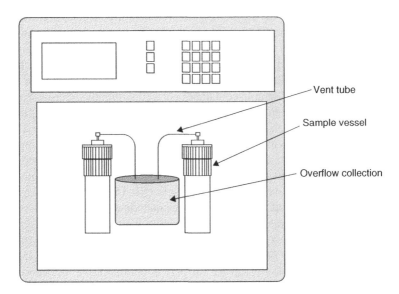

Figure 2.6 Schematic of a pressurized microwave digestion system.

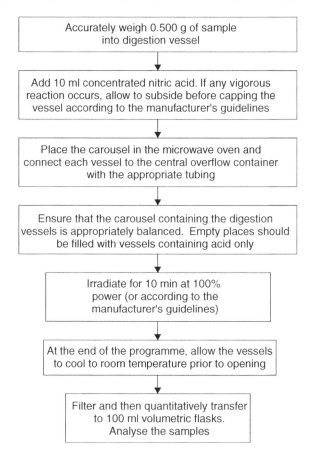

Figure 2.7 The United States Environmental Protection Agency (USEPA) procedure for the microwave digestion of sediments, sludges and soils [1, 5]. From Dean, J. R., *Methods for Environmental Trace Analysis*, AnTS Series. Copyright 2003. © John Wiley & Sons, Limited. Reproduced with permission.

that is located within the microwave cavity. Each vessel has a vessel body and an inner liner. The liner is made of 'TFM' fluoropolymer and has a volume of 100 ml. A patented safety system (AutoVent Plus™) allows the venting of excess pressure within each extraction vessel. The system works by lifting of the vessel cap to release excess pressure and then immediately resealing to prevent loss of sample. If solvent leaking from the extraction vessel(s) does occur, a solvent monitoring system will automatically shut off the magnetron while still allowing the exhaust fan to continue working. A typical procedure for the microwave digestion of sediments, sludges and soil samples is shown in Figure 2.7, while for siliceous and organically based matrices, see Figure 2.8.

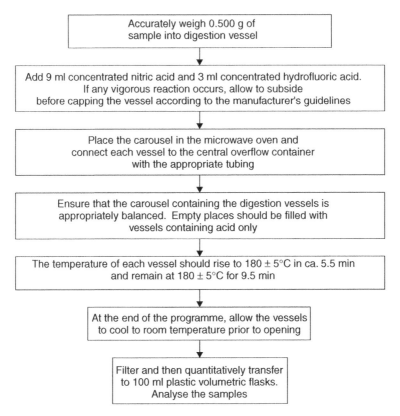

Figure 2.8 The United States Environmental Protection Agency (USEPA) procedure for the microwave digestion of siliceous and organically bound matrices [1, 6]. From Dean, J. R., *Methods for Environmental Trace Analysis*, AnTS Series. Copyright 2003. © John Wiley & Sons, Limited. Reproduced with permission.

2.3.3 Dry Ashing

Probably the simplest of all decomposition systems involves the heating of a sample in a silica or porcelain crucible in a muffle furnace at 400–800°C. The resultant residue can then be dissolved in mineral acid prior to ICP analysis. While this approach will undoubtedly destroy the organic matter of the sample matrix, it can also lead to loss of some volatile elements, including Hg, Pb, Cd, Ca, As, Sb, Cr and Cu. This can be partly remedied by the addition of compounds to retard the loss of volatiles; however, its analytical use is limited due to a number of severe disadvantages, namely, losses due to volatilization, resistance to ashing by some matrices, difficulties in dissolution of some ashed material, and the high risk of contamination.

2.3.4 Fusion

Some substances e.g. silicates and oxides, are not normally destroyed by the action of mineral acids. In this case, an alternative approach is required. *Fusion* involves the addition of an excess of reagent (10-fold) to the finely ground sample which is placed in a platinum crucible and then heated in a muffle furnace (300–1000°C). After heating for a specific period of time, a clear 'melt' should result, thus indicating completeness of the decomposition. After allowing to cool, the melt can be dissolved in mineral acid. Typical fusion reagents include the following:

- Sodium carbonate – 12–15 g of flux per g of sample are heated to 800°C and the resultant melt is then dissolved with HCl.

- Lithium meta- or tetraborate – a 10–20-fold excess of flux is heated to 900–1000°C and the resultant melt is then dissolved with HF.

- Potassium pyrosulfate – a 10–20-fold excess of flux is heated to 900°C and the resultant melt is then dissolved with H_2SO_4.

Note, however, that the addition of excess reagent (flux) can lead to a high risk of contamination, while, in addition, the high salt content of the final solution may lead to problems in the subsequent ICP analysis.

DQ 2.2

What problems might result from a high sample salt content?

Answer

A high salt content can cause problems in the ICP analysis step. For example, a high salt content can block the nebulizer used for sample introduction in inductively coupled plasma-based techniques.

2.4 Extraction Procedures

The term *extraction* is used instead of digestion as it refers to the removal of metals from sample matrices without destruction of the latter. This process of removing metals from sample matrices is often associated with the terms 'selective extraction' or *speciation*. The latter is often defined as 'the process of identifying and quantifying the different defined species, forms or phases present in a material' or 'the description of the amounts and types of these species, forms or phases present'. In some cases, it is possible to identify, by using single or sequential extractions, operationally defined determinations which identify 'groups' of metals without clear identification. In this case, it is possible to refer to, for example, ethylenediamine tetraacetic acid (EDTA)-extractable trace metals. The reasons why speciation is important is that metals (and metalloids) can be present in many forms, e.g. oxidation states, or as organometallic compounds, some of which are toxic. One approach to determine the speciation of metals (and metalloids) in environmental samples has been the linking of chromatographic separation with quantitation by ICP analysis. In this situation, the use of a suitable chromatographic technique, e.g. gas or liquid chromatography, is used to separate a metal complex prior to its detection by ICP analysis (see Section 3.5).

The use of single or sequential methods of extraction are required to remove metals, without altering their chemical form (speciation), from the sample matrix. This approach has not been notably accomplished in environmental studies, i.e. soil/sediment analysis. Extraction procedures have been developed which allow the isolation of a particular metal-containing soil phase, e.g. exchangeable component. The information can then be used to identify the potential likelihood of metal release, transformation, mobility or availability as the soils are exposed to weathering, pH changes, changes in land use, and their associated implications for environmental risk assessment. As the extraction methods are non-specific in nature, it has been necessary to define limits over which they can operate. The specifically defined soil phases are as follows:

- Water-soluble, soil solution, sediment pore water – this phase contains the most mobile and hence potentially available metals species.

- Exchangeable species – this phase contains weakly bound (electrostatically) metal species that can be released by ion-exchange with cations such as Ca^{2+}, Mg^{2+} or NH_4^+. Ammonium acetate is the preferred extractant as the complexing power of acetate prevents re-adsorption or precipitation of released metal ions. In addition, acetic acid dissolves the exchangeable species, as well as more tightly bound exchangeable forms.

- Organically bound – this phase contains metals bound to the humic material of soils. Sodium hypochlorite is used to oxidize the soil organic matter and release the bound metals. An alternative approach is to oxidize the organic matter with 30% hydrogen peroxide, acidified to pH 3, followed by extraction with ammonium acetate to prevent metal ion re-adsorption or precipitation.

- Carbonate bound – this phase contains metals that are dissolved by sodium acetate acidified to pH 5 with acetic acid.

- Oxides of manganese and iron – acidified hydroxylamine hydrochloride releases metals from the manganese oxide phase with minimal attack on the iron oxide phases. Amorphous and crystalline forms of iron oxides can be discriminated between by extracting with acid ammonium oxalate in the dark and under UV light, respectively.

The diversity and complexity of the approaches available have identified the major difficulties associated with producing suitable guidelines that would allow comparisons between different laboratories and different countries in assessing metal mobility in the soil environment. This has led to the development of single and sequential extraction procedures by the Standard, Measurements and Testing Programme (SM&T – formerly BCR) of the European Union (1987).

2.4.1 Single Extraction Procedures

Single extraction procedures based on the use of the following have been developed and applied in the analysis of soils and sediments:

(1) Ethylenediamine tetraacetic acid (EDTA) ($0.05\,\mathrm{mol}\,l^{-1}$)

(2) Acetic acid ($0.43\,\mathrm{mol}\,l^{-1}$)

(3) Diethylenetriamine pentaacetic acid (DTPA) ($0.005\,\mathrm{mol}\,l^{-1}$)

Examples of the procedures to be followed are shown in Figures 2.9–2.11 for EDTA, acetic acid and DTPA, respectively. All involve the addition of the reagent to the soil/sediment sample, followed by shaking for a specific time, centrifugation and then subsequent analysis of the supernatant liquid phase.

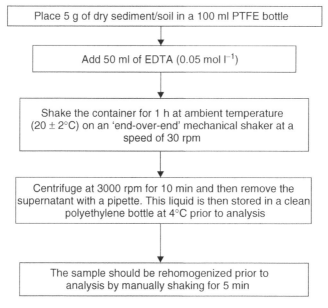

Figure 2.9 Procedure adopted in the single extraction method for metals, employing EDTA, as applied to the analysis of soils and sediments [1]. From Dean, J. R., *Methods for Environmental Trace Analysis*, AnTS Series. Copyright 2003. © John Wiley & Sons, Limited. Reproduced with permission.

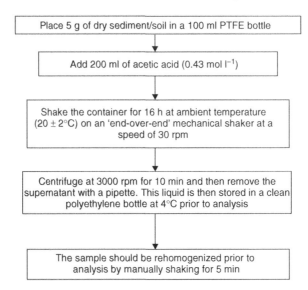

Figure 2.10 Procedure adopted in the single extraction method for metals, employing acetic acid, as applied to the analysis of soils and sediments [1]. From Dean, J. R., *Methods for Environmental Trace Analysis*, AnTS Series. Copyright 2003. © John Wiley & Sons, Limited. Reproduced with permission.

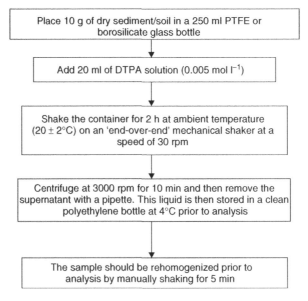

Figure 2.11 Procedure adopted in the single extraction method for metals, employing DTPA, as applied to the analysis of soils and sediments [1]. From Dean, J. R., *Methods for Environmental Trace Analysis*, AnTS Series. Copyright 2003. © John Wiley & Sons, Limited. Reproduced with permission.

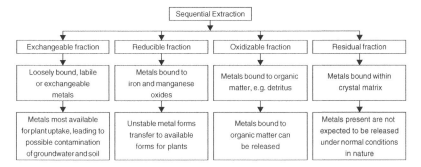

Figure 2.12 Overview of the sequential extraction method for metals, as applied to the analysis of soils and sediments [1]. From Dean, J. R., *Methods for Environmental Trace Analysis*, AnTS Series. Copyright 2003. © John Wiley & Sons, Limited. Reproduced with permission.

2.4.2 Sequential Extraction Procedures

The sequential extraction procedure consists of three (main) stages, plus a final (residual fraction) stage (Figure 2.12), as follows:

- *Step 1*. Metals extracted during this step are those which are exchangeable and in the acid-soluble fraction. This includes weakly absorbed metals retained on the sediment surface by relatively weak electrostatic interaction, metals that can be released by ion-exchange processes and metals that can be co-precipitated with carbonates present in many sediments. Changes in the ionic composition, influencing adsorption–desorption reactions, or lowering of pH, could cause mobilization of metals from such fractions. The experimental details are shown in Figure 2.13.

- *Step 2*. Metals bound to iron/manganese oxides are unstable under reducing conditions. Changes in the redox potential (E_h) could induce the dissolution of these oxides and could release adsorbed trace metals. The experimental details are shown in Figure 2.14.

- *Step 3*. Degradation of organic matter under oxidizing conditions can lead to a release of soluble trace metals bound to this component. Amounts of trace metals bound to sulfides might be extracted during this step. The experimental details are shown in Figure 2.15.

It is common to analyse for trace metals in the residual fraction. In this situation, the latter should contain naturally occurring minerals which may hold trace metals within their crystalline matrices. Such metals are not likely to be released under normal environmental conditions. The residual fraction is digested by using a 'pseudo-total' approach with aqua regia as most metal pollutants are not silicate-bound. However, for complete digestion, hydrofluoric acid is required.

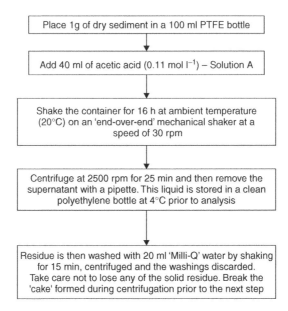

Figure 2.13 Details of Step 1 of the sequential extraction method (cf. Figure 2.12) [1]. From Dean, J. R., *Methods for Environmental Trace Analysis*, AnTS Series. Copyright 2003. © John Wiley & Sons, Limited. Reproduced with permission.

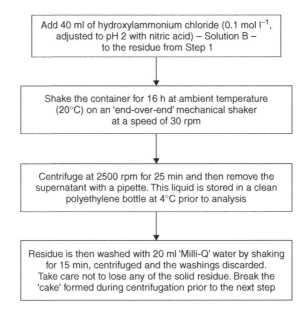

Figure 2.14 Details of Step 2 of the sequential extraction method (cf. Figure 2.12) [1]. From Dean, J. R., *Methods for Environmental Trace Analysis*, AnTS Series. Copyright 2003. © John Wiley & Sons, Limited. Reproduced with permission.

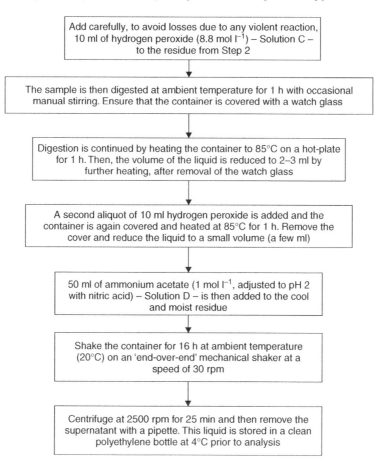

Figure 2.15 Details of Step 3 of the sequential extraction method (cf. Figure 2.12) [1]. From Dean, J. R., *Methods for Environmental Trace Analysis*, AnTS Series. Copyright 2003. © John Wiley & Sons, Limited. Reproduced with permission.

DQ 2.3

Why is the term 'pseudo-total' used?

Answer

Aqua regia is a good acid mixture for the digestion of soil and sediment samples. However, it cannot liberate from the matrix metal pollutants that are silicate-bound, i.e. part of the silicate 'backbone'. In this situation, if complete digestion is required then hydrofluoric acid must be used. As it is unlikely that the silicate-bound metals will leach from the soil or sediment, the use of aqua regia to give a 'pseudo-total' analysis is perfectively acceptable in this situation.

2.4.3 Enzymatic Digestion Procedures

An alternative strategy for releasing metals/metalloids from environmental and food matrices is the use of *enzymatic digestion*. This approach has the potential to specifically remove individual chemical forms from matrices, e.g. speciation.

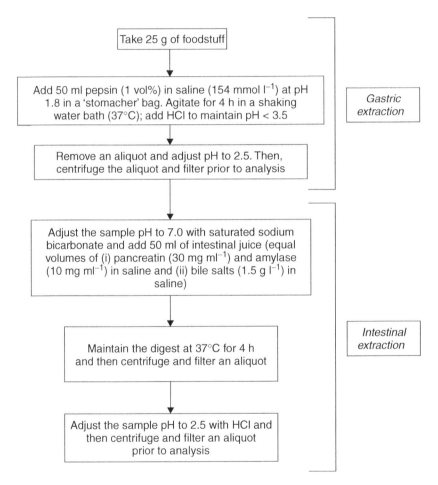

Figure 2.16 The gastro-intestinal extraction procedure used to investigate the speciation of metals from ingested foodstuffs [1, 8]. From Dean, J. R., *Methods for Environmental Trace Analysis*, AnTS Series. Copyright 2003. © John Wiley & Sons, Limited. Reproduced with permission.

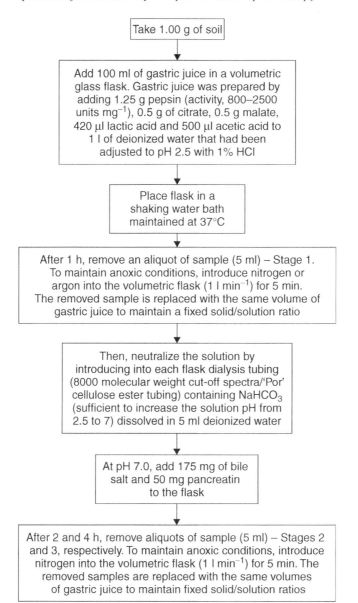

Figure 2.17 The physiologically based extraction test (PBET) used to determine the bioaccessibilty of metals from soil [1, 9]. From Dean, J. R., *Methods for Environmental Trace Analysis*, AnTS Series. Copyright 2003. © John Wiley & Sons, Limited. Reproduced with permission.

The procedure is based on the use of selected enzymes, often hydrolytic enzymes, to selectively remove metals/metalloids from sample matrices under mild conditions of temperature and pH, thus allowing the potential for the metal/metalloid to retain its original chemical form (or at least be similar). Typical protolytic enzymes used for this process include the following [7]:

• Lipases – hydrolyse fats into long-chain fatty acids and glycerol.

• Amylases – hydrolyse starch and glycogen to maltose and to residual polysaccharides.

• Proteases – attack the peptidic bonds of proteins and peptides.

Examples of the use of this approach for the removal of metals/metalloids from foodstuffs and soil are highlighted in Figures 2.16 and 2.17, respectively.

Summary

A variety of methods for the determination of total metal in aqueous and solid samples have been described. While the most common method of sample preparation for solid samples is acid-digestion, other approaches, most notably using extraction procedures, have been described, including enzymatic digestion. In contrast, the most common method of sample preparation for aqueous samples is liquid–liquid extraction, although other techniques have been used, in particular, ion-exchange which allows the possibility of 'on-line' coupling to an inductively coupled plasma source.

References

1. Dean, J. R., *Methods for Environmental Trace Analysis*, AnTS Series, Wiley, Chichester, UK, 2003.
2. Kirkbright, G. F. and Sargent, M., *Atomic Absorption and Fluorescence Spectroscopy*, Academic Press, London, 1974.
3. United States Environmental Protection Agency, 'Acid digestion of sediments, sludges and soils', *EPA Method 350B*, National Technical Information Services, Springfield, VA, USA, 1996.
4. Abu-Samra, A., Morris, J. S. and Koityohann, S. R., *Anal. Chem.*, **47**, 1475–1477 (1975).
5. United States Environmental Protection Agency, 'Microwave-assisted acid digestion of sediments, sludges and soils', *EPA Method 3051*, National Technical Information Services, Springfield, VA, USA, 1994.
6. United States Environmental Protection Agency, 'Microwave-assisted acid digestion of siliceous and organically based matrices', *EPA Method 3052*, National Technical Information Services, Springfield, VA, USA, 1996.
7. Bermejo, P., Capelo, J. L., Mota, A., Madrid, Y. and Camara, C., *Trends Anal. Chem.*, **23**, 654–663 (2004).
8. Crews, H. M., Burrell, J. A. and McWeeny, D. J., *Lebensm. Unters Forsch.*, **180**, 221–226 (1985).
9. Ruby, M. V., Davies, A., Schoof, R., Eberle, S. and Sellstone, C. M., *Environ. Sci. Technol.*, **30**, 422–430 (1996).

Chapter 3

Sample Introduction Procedures for Inductively Coupled Plasmas

Learning Objectives

- To be aware of the different forms of sample introduction devices used for inductively coupled plasmas.
- To understand the principle of operation of a range of nebulizers.
- To be able to prevent blockages or unblock a nebulizer.
- To appreciate the importance of a spray chamber and its use with a nebulizer.
- To be aware of the scope and operation of discrete sample introduction devices used for ICPs.
- To appreciate the different calibration strategies used for laser ablation.
- To appreciate the importance of continuous sample introduction devices used for ICPs.
- To be able to design a simple flow injection system.
- To understand the principle of coupling high performance liquid chromatography and gas chromatography systems to ICPs.
- To appreciate the advantages and limitations offered by hydride generation and cold vapour techniques for ICPs.

3.1 Introduction

The efficient introduction of a sample into an inductively coupled plasma (ICP) is crucial for the detection of elements of different concentrations. However,

Practical Inductively Coupled Plasma Spectroscopy J. R. Dean
© 2005 John Wiley & Sons, Ltd

while sample introduction has been one of the most investigated areas of ICP technology, the limitations in its effectiveness are still evident. For liquid samples, the combination of a nebulizer and spray chamber is the most common approach, but results in only a small portion (often < 2%) of the sample reaching the ICP. While other approaches for the introduction of solid and liquid samples are possible, the additional complexity associated with their use has often precluded their widespread acceptance for routine analysis. This chapter considers the different approaches possible for the introduction of both liquid and solid samples into an ICP.

3.2 Nebulizers

The most common nebulizer in use today is the **pneumatic concentric** glass nebulizer (Figure 3.1). The design of the nebulizer is based on the work of Gouy [2], but was first fabricated in 1979 by Meinhard [3]. It operates by using the Venturi effect principle. Essentially, argon gas introduced in the side arm is able to exit at the nozzle, so causing the development of a region of low pressure. This results in liquid sample being drawn up through the capillary tube and exiting through the nozzle. The physical interaction of argon gas and liquid sample causes a coarse aerosol to be produced.

DQ 3.1
Why not simply introduce this coarse aerosol into the ICP?

Answer
The introduction of the aerosol directly into the ICP would either (a) cause the ICP to be extinguished, or (b) lead to severe interferences due to the lowering in temperature of the plasma (see also Section 4.1). Therefore, the coarse aerosol is then subjected to further treatment in a **spray chamber**.

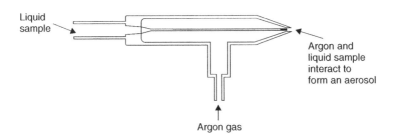

Figure 3.1 Schematic diagram of the pneumatic concentric nebulizer [1]. From Dean, J. R., *Atomic Absorption and Plasma Spectroscopy*, 2nd Edition, ACOL Series, Wiley, Chichester, UK, 1997. © University of Greenwich, and reproduced by permission of the University of Greenwich.

The pneumatic concentric nebulizer can become blocked during use. To prevent or minimize the chance of this happening, the following steps should be taken:

(1) **Filter the carrier gas.** The use of 'in-line' gas filters can reduce/eliminate particles accidentally introduced into the gas supply by, for example, the use of PTFE tape, as used as part of the gas line plumbing process.

(2) **Filter the sample.** Filtering the sample through a $0.2\,\mu$m filter will reduce particulates that may build up in the glass capillary, so leading to reduced liquid flow.

(3) **Rinse the nebulizer.** After completing all analyses, it is important to rinse the nebulizer prior to turning off the argon gas supply. If a low-pH sample has been analysed, rinse the nebulizer with a low-pH rinse solution; conversely, for a high-pH sample use a high-pH rinse solution. For organic samples, use an organic solvent to rinse the nebulizer. Finally, rinse the nebulizer with 'ultrapure' water prior to switching off the argon gas supply.

If the nebulizer does become blocked, the following cleaning procedures can be used:

(1) **Remove the nebulizer from its mounting** and visually examine under $20\times$ or $30\times$ magnification.

(2) **Particles wedged inside the nebulizer.**

   (a) Carefully and gently tap the liquid input of the nebulizer. If the particles come loose, then repeat the tapping to allow the particle to progress to the exit orifice.

   (b) Apply compressed gas (up to $30\,$psi (20.7×10^4 Pa)) to the nozzle.

   (c) Backflush the nozzle by using isopropyl alcohol.

(3) **Solid material present, but flow through nebulizer is still possible.**

   (a) Inject an appropriate solvent into the nozzle to dissolve the solid deposit and then remove the solvent with compressed gas.

   (b) Repeat step (a), but gently heat the nebulizer.

(4) **Solid material present and nebulizer is blocked.** Gently heat the nebulizer at the point of the blockage and then carefully apply gas pressure at the sample input tube.

(5) **Nozzle is encrusted with crystalline deposits.**

   (a) Immerse the nozzle in an appropriate rinse solution.

   (b) Apply compressed gas (up to $30\,$psi (20.7×10^4 Pa)) to the nozzle.

(6) **No foreign matter is visible.** Immerse the nozzle in hot, concentrated nitric acid and repeat as necessary.

In **cross-flow** nebulizers, the liquid sample and argon gas interact at perpendicular to one another. The original **cross-flow** nebulizer consists of two capillary needles positioned at 90° to each other with their tips not quite touching (Figure 3.2). The argon carrier gas flows through one capillary tube while through the other capillary the liquid sample is pumped. At the exit point, the force of the escaping carrier gas is sufficient to shatter the sample into a coarse aerosol. Modifications in this design have been used for sample solutions with higher dissolved solids content e.g. the **V-groove** or Babington-type nebulizer. This type of nebulizer is designed to generate coarse aerosols from aqueous samples with high-solids contents (up to 20%). The design of this nebulizer (Figure 3.3) allows the sample solution to be pumped along a V-grooved channel. Midway along the channel is a small orifice through which the carrier gas can escape. As the sample passes over the orifice, the escaping argon gas causes generation of a coarse aerosol.

In **ultrasonic** nebulizers, the sample solution is pumped onto a vibrating piezoelectric transducer, operating at between 200 kHz and 10 MHz. The action of the vibrating crystal is sufficient to transform the liquid sample into an aerosol. The latter is then transported by the argon carrier gas through a heated tube and then a condenser (Figure 3.4). This has the effect of removing the solvent. Under such conditions, the aerosol is desolvated and reaches the plasma source in a fine, dry state. The major advantage of the ultrasonic nebulizer is its increased transport efficiency – of the order of 10% (compared with the pneumatic nebulizer). For

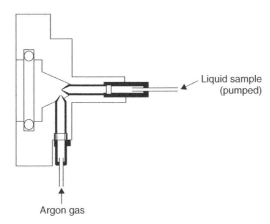

Figure 3.2 Schematic diagram of the cross-flow nebulizer [1]. From Dean, J. R., *Atomic Absorption and Plasma Spectroscopy*, 2nd Edition, ACOL Series, Wiley, Chichester, UK, 1997. © University of Greenwich, and reproduced by permission of the University of Greenwich.

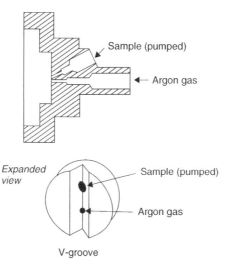

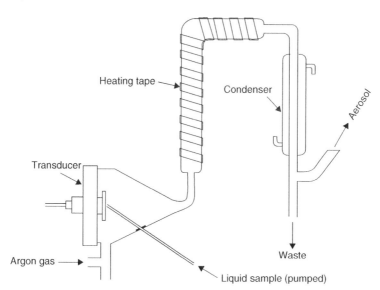

Figure 3.3 Schematic diagram of the V-groove, high-solids nebulizer [1]. From Dean, J. R., *Atomic Absorption and Plasma Spectroscopy*, 2nd Edition, ACOL Series, Wiley, Chichester, UK, 1997. © University of Greenwich, and reproduced by permission of the University of Greenwich.

Figure 3.4 Schematic diagram of the ultrasonic nebulizer [1]. From Dean, J. R., *Atomic Absorption and Plasma Spectroscopy*, 2nd Edition, ACOL Series, Wiley, Chichester, UK, 1997. © University of Greenwich, and reproduced by permission of the University of Greenwich.

further information on the fundamentals of nebulizers, the reader is directed to a comprehensive review published in 1988 on their use in ICP spectrometry [4].

3.3 Spray Chambers and Desolvation Systems

Introduction of the coarse aerosols generated by the nebulizers directly into the plasma source would extinguish or induce cooling of the plasma, hence leading to severe matrix interferences. The inclusion of a spray chamber (see Figure 3.5) can result in the production of a more appropriate aerosol for the plasma source.

DQ 3.2
What effects can a spray chamber offer?

Answer
These are as follows:

(1) reduce the amount of aerosol reaching the plasma;

(2) decrease the turbulence associated with the nebulization process;

(3) reduce the aerosol particle size.

It has been determined that the ideal particle size for the plasma processes to occur, i.e. atomization, followed by either excitation/emission or ionization, is around $10\,\mu m$. Several spray chamber designs are available, including the following:

• double-pass or Scott-type
• cyclonic

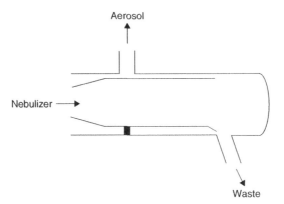

Figure 3.5 Schematic diagram of a double-pass spray chamber (Scott-type) [1]. From Dean, J. R., *Atomic Absorption and Plasma Spectroscopy*, 2nd Edition, ACOL Series, Wiley, Chichester, UK, 1997. © University of Greenwich, and reproduced by permission of the University of Greenwich.

- single-pass, direct or cylindrical type

An ideal spray chamber should have all of the following features:

- a large surface area to induce collisions and fragmentation of the coarse aerosol
- minimal dead volume to prevent dilution of the sample
- easy removal of condensed sample to waste without inducing pressure pulsing

The most common type of spray chamber is the **double-pass**. This is comprised of two concentric tubes, an inlet for the nebulizer, an exit for the finer aerosol and a waste drain (Figure 3.5). The double-pass spray chamber is positioned so as to allow excess liquid (aerosol condensation) to flow to waste. The nebulizer-generated aerosol is introduced into the inner tube and exits having reversed its direction (180°) into the ICP torch. Interaction of the coarse aerosol with the internal surfaces of the double-pass spray chamber leads to the production of a finer aerosol (with consequent excess liquid going to waste). This design of spray chamber also acts to reduce the turbulence of the nebulizer-generated aerosol, hence leading ultimately to greater signal stability. A drawback of this design of spray chamber is the presence of so-called 'dead volumes' that are not easily reached by the argon nebulizer gas. This leads to an increased 'wash-out' time, i.e. the time taken to remove all existing sample prior to the introduction of the next sample. Wash-out time is normally measured as the time taken for the ICP system to measure 1% of the steady-state signal (Figure 3.6).

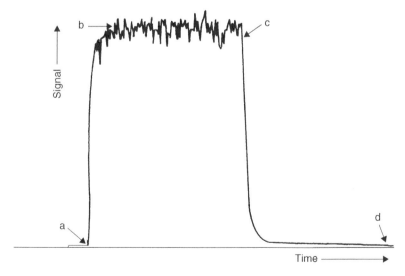

Figure 3.6 A typical time–signal profile for ICP analysis: (a) sample introduced via nebulizer/spray chamber; (b) start of steady-state signal; (c) loss of steady-state signal commences; (d) signal decreases to 1% of the steady-state signal.

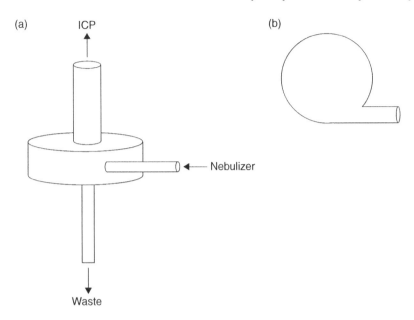

Figure 3.7 Schematic diagram of the cyclonic spray chamber: (a) side view; (b) top view.

In the **cyclonic** spray chamber (Figure 3.7), the aerosol is introduced tangentially to induce swirling. The initial process involves the aerosol swirling downwards close to the spray chamber wall. At the bottom of the spray chamber, a second inner spiral carries the aerosol to the exit point. The combined process of induced tangential flow and subsequent aerosol swirling leads to a reduction in aerosol particle size. The **single-pass**, direct or cylindrical spray chamber (Figure 3.8) often contains an impact bead for aerosol production. It is recommended that this type of spray chamber is connected to a desolvation system to reduce sample load into the plasma.

Desolvation systems are designed to reduce the solvent, i.e. water or volatile solvents, loading into the plasma. In its simplest form, it consists of a spray chamber fitted with a thermostated water jacket.

An interesting development in the introduction of liquid samples into plasmas has been the use of **micro-nebulizers**. These devices have been developed for cases where sample is limited, in, for example, clinical samples. The most common approach to this has been to reduce the nebulizer gas flow rate from the conventional 1 ml min^{-1} to the µl min^{-1} level. Various pneumatic concentric micro-nebulizers have been developed and include the high-efficiency nebulizer and direct-injection nebulizer. For further information on the fundamentals of spray chambers, the reader is directed to a comprehensive review published in 1988 on their use in ICP spectrometry [5].

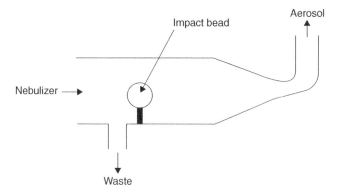

Figure 3.8 Schematic diagram of a single-pass spray chamber (direct or cylindrical type [1]. From Dean, J. R., *Atomic Absorption and Plasma Spectroscopy*, 2nd Edition, ACOL Series, Wiley, Chichester, UK, 1997. © University of Greenwich, and reproduced by permission of the University of Greenwich.

3.4 Discrete Sample Introduction

The direct introduction of a sample into a plasma source can be achieved by using either the argon carrier gas or a liquid carrier stream. Discrete sample introduction has the advantage of presenting the plasma source with all of the analyte in a short time. Therefore, while the residence time of the analyte within the plasma is short, all of the analyte is available for analysis, thus leading to improved analyte sensitivity.

The principle of **electrothermal vaporization** (ETV) is that the aqueous sample is pipetted onto a graphite surface (Figure 3.9) and heated at different rates. The temperature of the graphite surface can be controlled to preferentially allow destruction and removal of the sample matrix (ashing), but not the analyte of interest. Obviously, this may not be possible and the addition of matrix modifiers (substances that allow the formation of more stable compounds) may be required to prevent analyte loss. After ashing, the temperature is rapidly increased to allow vaporization of the analyte directly into the plasma source.

DQ 3.3

Why is vaporization of the analytes required instead of atomization?

Answer

The formation of reactive atoms is undesirable as they can interact with surfaces in which they come into contact and be more easily lost. Vaporization of the analytes is therefore preferred.

An additional heating stage may be required for cleaning of residual material from the graphite surface, prior to cooling. A typical heating process from

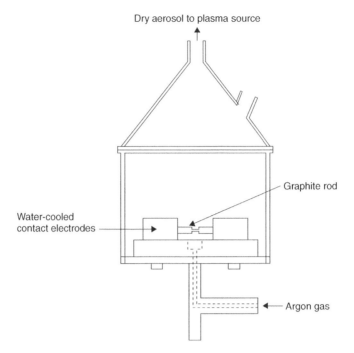

Figure 3.9 Schematic representation of electrothermal vaporization (graphite-rod type) [1]. From Dean, J. R., *Atomic Absorption and Plasma Spectroscopy*, 2nd Edition, ACOL Series, Wiley, Chichester, UK, 1997. © University of Greenwich, and reproduced by permission of the University of Greenwich.

injection of the sample to the cooling stage may last up to 60 s. This sample introduction technique has the advantage of being able to introduce small samples directly to the ICP with high transport efficiency.

Laser ablation allows a solid sample to be mobilized (ablated) and transported direct to the plasma torch via the argon carrier gas. Laser ablation has the following benefits:

• applicable to any solid sample

• no sample size requirements

• no sample preparation, no reagents or solution waste

• spatial characterization information available

However, it does have some limitations, as follows:

• amount of sample ablated is dependent upon laser and sample properties

• 'matrix-matched' standards required for calibration

• fractionation occurs for samples containing low-melting-point elements

One of the most common lasers used for ablation of samples is the Nd:YAG which operates in the near infrared at 1064 nm.

SAQ 3.1
One of the most commonly used lasers is the Nd:YAG. What does Nd:YAG stand for?

With optical frequency doubling, tripling, quadrupling and quintupling, the Nd:YAG laser can also be operated at the following wavelengths 532, 355, 266 and 213 nm, respectively. In addition, *excimer* lasers have been used. As the operating wavelengths are dependent upon the gas, the choice of output wavelengths are 308, 248, 193 and 157 nm for XeCl, KrF, ArF and F_2, respectively.

A typical experimental layout for laser ablation coupled to a plasma source consists of an ablation chamber, a lens to allow focusing, an adjustable platform for positioning in the *x*-, *y*- and *z*-directions, and a charge-coupled device (CCD) camera for remote viewing of the sample surface (Figure 3.10). The sample is placed inside the ablation chamber, which is fitted with a fused silica window. The chamber is then flushed with argon carrier gas to transport the ablated sample material directly to the ICP torch.

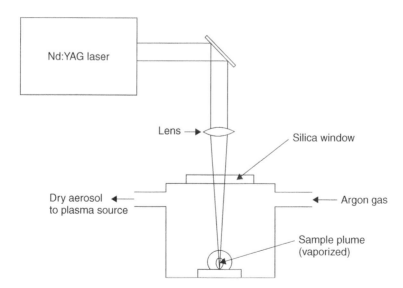

Figure 3.10 Schematic representation of laser ablation [1]. From Dean, J. R., *Atomic Absorption and Plasma Spectroscopy*, 2nd Edition, ACOL Series, Wiley, Chichester, UK, 1997. © University of Greenwich, and reproduced by permission of the University of Greenwich.

The typical size of the crater generated when using a Nd:YAG laser is 10–100 μm in diameter. Ablation of a hard material, e.g. stainless steel, results in a crater profile with upraised walls (Figure 3.11(a)) while the ablation of a softer material, e.g. a semiconductor, will undergo melting and subsequent flowing away from the hot central position (Figure 3.11(b)).

DQ 3.4

Identify two ways in which sample losses may occur in laser ablation.

Answer

(1) 'Hot' sample ejected from the surface can be cooled by the argon carrier gas, so leading to redeposition.

(2) The ejected material can be lost enroute *to the plasma source by deposition in the connecting tubing.*

A major disadvantage of laser ablation is the difficulty in obtaining samples for instrument calibration. However, three different calibration strategies exist, as follows:

- matrix-matched direct solid ablation

- dual introduction (sample-standard)

- direct liquid ablation

Matrix-matched direct solid ablation using external calibration is the most common approach for laser ablation–ICP analysis. It is essential to use matrix-matched standards as the quantity of material ablated per laser pulse is sample matrix-dependant (cf. the differences between hard and soft materials). The availability of certified reference materials (CRMs) allows matrix-matched standards to be used. Certified reference materials (see Section 1.8) exist for a range of solid matrices, including steels, alloys, glass and ceramics. Several approaches are possible to produce matrix-matched standards, including the 'bricketing' of powdered material, i.e. compression using a hydraulic press in the presence of a binder, addition of standard solutions to a powdered matrix, co-precipitation of the element of interest (see Section 2.2.3) and fusion techniques (see Section 2.3.4).

In a **dual introduction (sample-standard)** system, laser-ablated material and solution-nebulized standards are both introduced into the ICP in a sequential manner. This allows signals from the ablated material to be compared directly with solution standards introduced via a nebulizer/spray chamber arrangement. While this approach is an attractive alternative, particularly when CRMs are difficult to obtain or simply not available, it does suffer from a major disadvantage. The drawback is the different analyte responses from the two methods due to the

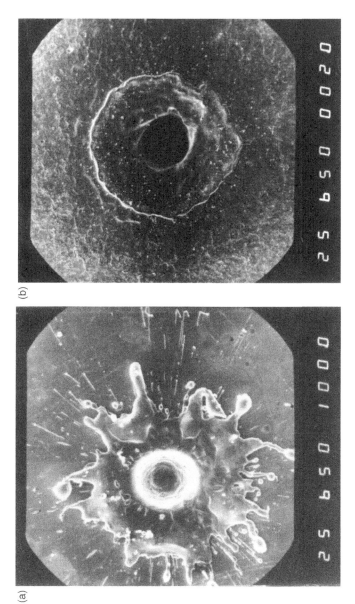

Figure 3.11 Effect of a Nd:YAG laser on the surfaces of (a) stainless steel and (b) a semiconductor material (cadmium mercury telluride).

introduction of a 'dry' sample from laser ablation and a 'wet' aerosol from the nebulization process. For example, in ICP–MS the use of laser ablation and hence a 'dry' sample introduction system leads to reduced oxide interferences due to the elimination of water from the sample. This disadvantage can be remedied, in part, by the use of a desolvation system attached to the nebulizer/spray chamber.

An alternative calibration strategy uses **direct liquid ablation** to introduce aqueous standards into the plasma. To improve the optical absorption characteristics of the standard solution, it may be necessary to add a *chromophore*. The presence of a chromophore may lead to improved laser energy coupling to the aqueous solution standard in such a manner as to allow the ablation process to occur within the surface layers of the liquid, thereby leading to the production of a fine aerosol. The ideal characteristics of a chromophore include the following:

- an ability to absorb strongly at the laser wavelength

- does not lead to precipitation in the standard solution

- is non-toxic

The major advantage of this approach is that it offers a new surface for each subsequent laser pulse.

For qualitative work, the limited sampling size can be very useful for characterizing impurities in manufactured goods or for mineral identification in geological samples.

3.5 Continuous Sample Introduction

Flow injection (FI) and chromatography are both methods of introducing aqueous samples in a flowing stream into the plasma source. **Flow injection** is a 'multi-method' system that allows the user to create unique sample presentation facilities for the plasma source. In principle, any FI system will consist of a peristaltic pump, injection valve, 'sample alteration/modification' component, and an interface to the nebulizer. The mode of 'sample alteration/modification' is diverse and can be developed according to the particular application.

DQ 3.5

Can you think of any possibility for calibrating a system by using this methodology?

Answer

In principle, it would be possible to have a carrier stream in which the sample is flowing being mixed with a carrier stream into which different

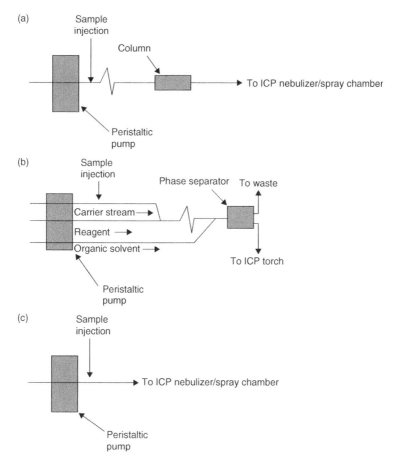

Figure 3.12 Components of a 'sample alteration/modification' module: (a) a low-pressure chromatography column, used for retention of the analyte in preference to the sample matrix; (b) an 'on-line' solvent extraction system; (c) a means of delivering a small discrete sample with the minimum of dilution to the nebulizer/spray chamber.

standard solutions could be introduced. This would provide a method for generating a standard additions calibration plot 'on-line'.

Typically, however, the 'sample alteration/modification' module may consist of (i) a low-pressure chromatography column used for retention of the analyte in preference to the sample matrix (Figure 3.12(a)), (ii) a gas–liquid interface for separation of a hydride-generated species (see Section 3.6), (iii) an 'on-line' solvent extraction system (Figure 3.12(b)), or (iv) a means of delivering a small discrete sample with the minimum of dilution to the nebulizer/spray chamber (Figure 3.12(c)).

Chromatography can be useful for two reasons, namely (i) elemental species information may be obtained – so-called speciation studies (see Section 2.4) – and (ii) potential matrix interferences can be separated. The type of chromatography used largely depends on the nature of the analyte to be separated but is broadly based on high performance liquid chromatography (HPLC), gas chromatography (GC) or supercritical fluid chromatography (SFC). Within these specific chromatography areas, particular variations can be used, e.g. HPLC includes ion-exchange, reversed-phase and size-exclusion chromatography.

Interfacing an HPLC system with an ICP normally offers few challenges with respect to physical coupling of the two techniques. Liquid chromatography flow rates are often of the order of $1 \, ml \, min^{-1}$, which is directly compatible with the typical aspiration rates of a conventional nebulizer/spray chamber system for ICP analysis. Therefore, the simple connection of the output from the HPLC system to the input of the nebulizer via the minimum amount of polytetrafluoroethylene (PTFE) or poly(ether ether ketone) (PEEK) tubing is all that is required, with the overall process being limited to the typical inefficiencies of the nebulizer/spray chamber system (1–2% transport efficiency). The composition of the mobile phase in HPLC can also lead to problems in terms of isobaric interferences (in MS) due to the introduction of organic solvents and high-salt-content (buffer) solutions.

Similarly, the interfacing of a GC system with an ICP does not present any difficult challenges. As GC separates volatile compounds, it is essential in the interface to maintain the analytes in the vapour state. This is often achieved via a heated transfer line (200–250°C) which allows the carrier gas (argon, helium or nitrogen) and volatile compounds to be directly transported to the ICP torch (no nebulizer/spray chamber required). This not only allows a much higher transport efficiency than in HPLC but leads to enhanced sensitivities of the elements being investigated.

3.6 Hydride and Cold Vapour Techniques

Hydride generation is a gas-phase sample introduction technique that allows the determination of ultra-trace and trace levels of analytes to be measured. It is limited to a number of elements that are capable of forming volatile hydrides at ambient temperature, e.g. arsenic, antimony, bismuth, selenium, tellurium and tin. Under acid conditions, and in the presence of a reducing agent, e.g. sodium tetraborohydride, covalent hydrides are formed, e.g. AsH_3, SbH_3, BiH_3, H_2Se, H_2Te and SnH_4.

The principle of hydride generation can be described in four steps, as follows:

(1) chemical generation of the hydrides;

(2) collection and pre-concentration of the evolved hydrides (if necessary);

(3) transport of the hydrides and gaseous byproducts to the ICP;

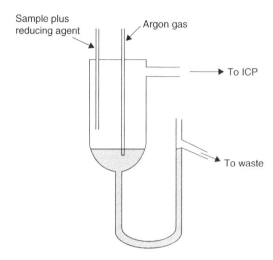

Figure 3.13 An example of a gas–liquid separation device for hydride generation.

(4) atomization of the hydrides, followed by either atomization/emission in AES or ionization in MS.

As an example, Equation (3.1) describes the chemical generation of the arsine hydride (AsH_3):

$$3BH_4^- + 3H^+ + 4H_3AsO_3 \longrightarrow 3H_3BO_3 + 4\mathbf{AsH_3} + 3H_2O \qquad (3.1)$$

However, when the basic borohydride is added to an acidic solution, excess hydrogen is liberated:

$$BH_4^- + 3H_2O + H^+ \longrightarrow H_3BO_3 + 4\mathbf{H_2} \qquad (3.2)$$

Various gas–liquid separation devices have been used for continuous, flow-injection or batch-mode separation. An example of such a separation device is shown in Figure 3.13.

Cold vapour generation is exclusively reserved for the element mercury. In this situation, the mercury present in the sample is reduced, usually with tin(II) chloride, to elemental mercury:

$$Sn^{2+} + Hg^{2+} \longrightarrow Sn^{4+} + Hg^0 \qquad (3.3)$$

The mercury vapour generated is then transported to the ICP by argon carrier gas.

Summary

Sample introduction is key to the efficiency of atomization/ionization in an inductively coupled plasma. The main approach to sample introduction is via a nebulizer/spray chamber arrangement. The range of possibilities in terms of choice of nebulizer and design of the spray chamber are discussed in this chapter. Alternative sample introduction techniques are highlighted and described – these include electrochemical vaporization, laser ablation, hydride and cold-vapour generation, and the use of flow-injection technologies.

References

1. Dean, J. R., *Atomic Absorption and Plasma Spectroscopy*, 2nd Edition, ACOL Series, Wiley, Chichester, UK, 1997.
2. Gouy, M., *An. Chim. Phys.*, **XVIII**, 5–101 (1879).
3. Meinhard, J. E., 'Pneumatic nebulizers, present and future', in *Applications of Plasma Emission Spectrochemistry*, Barnes, R. M. (Ed.), Heyden & Son, Inc., Philadelphia, PA, USA, 1979, pp. 1–14.
4. Sharp, B. L., *J. Anal. At. Spectrom.*, **3**, 613–652 (1988).
5. Sharp, B. L., *J. Anal. At. Spectrom.*, **3**, 939–963 (1988).

Chapter 4

The Inductively Coupled Plasma and Other Sources

Learning Objectives

- To be aware of the different forms of plasma sources available.
- To understand and explain the principle of operation of each plasma source.
- To appreciate the concept of temperature in relation to plasmas.
- To appreciate the role each plasma source has in atomic spectroscopy.

4.1 Introduction

A plasma is the co-existence, in a confined space, of the positive ions, electrons and neutral species of an inert gas, typically argon or helium. The most common plasma sources used in atomic spectroscopy are the (radiofrequency) inductively coupled plasma (ICP), direct-current plasma (DCP), microwave-induced plasma (MIP) and glow discharge. All plasmas can be characterized by their temperature. However, the concept of plasma temperature is difficult to answer without further investigation.

DQ 4.1

How do you measure temperature in the laboratory or at home?

Answer

In the laboratory or at home, temperature is measured by using a thermometer. However, in a plasma source this is not possible. Even though

Practical Inductively Coupled Plasma Spectroscopy   J. R. Dean
© 2005 John Wiley & Sons, Ltd

*the plasma is electrically neutral, it is **not** in thermodynamic equilibrium. Hence, it is not possible to characterize a single temperature. Four temperatures can be used to characterize the plasma, i.e. excitation, ionization, electron and gas temperatures. The **excitation temperature** is a measure of the population density of energy levels (see Section 5.1.2), while the **ionization temperature** represents the population density of different ionization states, the **electron temperature** the kinetic energy of the electrons, and the **gas temperature** the kinetic energy of the atoms. In each case, the typically quoted ICP temperatures range from 7000 to 10 000 K. Unfortunately, these are well above the range covered by the common form of temperature measurement, i.e. a thermometer. Thus, in order to measure the temperature of a plasma source the scientist must use alternative approaches, such as those based on spectroscopic methods. The **excitation temperature**, for example, can be measured by means of the Boltzmann equation (see Section 5.1.2). The **plasma temperature** is also **inhomogenous**, in that temperature variation occurs both radially and axially. This means that plasmas are complex sources. Nevertheless, it is their analytical use that is of interest in this present book.*

4.2 Inductively Coupled Plasma

The ICP is formed within the confines of three concentric glass tubes of a plasma torch (Figure 4.1). Each concentric glass tube has an entry point, with the intermediate (plasma) and external (coolant) tubes having tangentially arranged entry points and the inner tube consisting of a capillary tube through which the aerosol is introduced from the nebulization/spray chamber. Located around the outer glass tube is a coil of copper tubing through which water is recirculated. Power input to the ICP is achieved through this copper (load or induction) coil, typically in the range $0.5-1.5$ kW at a frequency of 27 or 40 MHz. The inputed power causes the induction of an oscillating magnetic field whose lines of force are axially orientated inside the plasma torch and follow eliptical paths outside the induction coil (Figure 4.2). At this point in time, no plasma exists. In order to initiate the plasma, the carrier gas flow is first switched off and a spark added momentarily from a Tesla coil, which is attached to the outside of the plasma torch by means of a piece of copper wire. Instantaneously, the spark, a source of 'seed' electrons, causes ionization of the argon carrier gas. This process is self-sustaining so that argon, argon ions and electrons co-exist within the confines of the plasma torch but protruding from the top in the shape of a bright white luminous 'bullet'. The escaping high-velocity argon gas, causing air entrainment back towards the plasma torch itself, forms the characteristic 'bullet shape' of the ICP. In order to introduce the sample aerosol into the confines of the hot plasma gas $(7000-10\,000$ K$)$, the carrier gas is switched on – this punches a hole in the

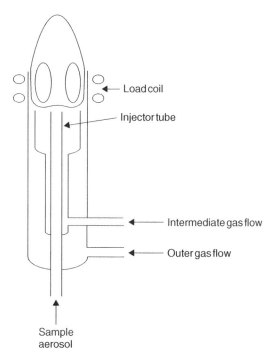

Figure 4.1 Schematic diagram of an inductively coupled plasma torch [1]. From Dean, J. R., *Atomic Absorption and Plasma Spectroscopy*, 2nd Edition, ACOL Series, Wiley, Chichester, UK, 1997. © University of Greenwich, and reproduced by permission of the University of Greenwich.

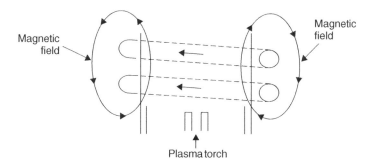

Figure 4.2 Schematic representation of the formation of an inductively coupled plasma [1]. From Dean, J. R., *Atomic Absorption and Plasma Spectroscopy*, 2nd Edition, ACOL Series, Wiley, Chichester, UK, 1997. © University of Greenwich, and reproduced by permission of the University of Greenwich.

centre of the plasma, thus creating the characteristic 'doughnut' or toroidal shape of the ICP. In the conventional ICP system, the emitted radiation is viewed laterally, or side-on. Therefore, the element radiation of interest is 'viewed' through the luminous plasma.

DQ 4.2

Can you identify any potential difficulty in viewing the plasma side-on through the luminous plasma? Can you suggest an alternative viewing position?

Answer

Figure 4.3 compares the typical background emission characteristics observed for a conventional, side-on viewed plasma with that of an axially viewed plasma. The two major features of the spectra are the presence of a large number of emission lines and a background continuum. The emission lines are mainly due to the source gas, namely argon, but also the presence of atmospheric gases, e.g. nitrogen, and the breakdown components of water, e.g. OH. It can be seen that the side-on viewed plasma has a higher background continuum than the axially viewed plasma. In either case, the background continuum is due to radiative recombination of electrons and ions ($M^+ + e^- \rightarrow M + h\nu$) and the radiation loss of energy by accelerated electrons (Bremsstralung radiation). Commercial instruments are now available that allow axial viewing of the plasma.

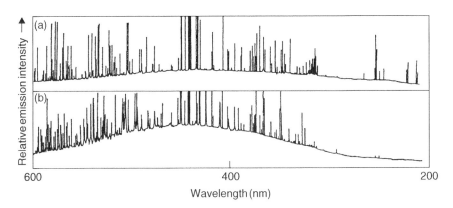

Figure 4.3 Comparison of the spectral features and background emission characteristics observed for (a) an axially viewed ICP, and (b) a conventional, side-on viewed ICP [2]. From Davies, R., Dean, J. R. and Snook, R. D., *Analyst*, **110**, 535–540 (1985). Reproduced by permission of The Royal Society of Chemistry.

4.3 Direct-Current Plasma

The direct-current plasma (DCP) is an electrical discharge struck between two anodes and a cathode. The arrangement of the electrodes (Figure 4.4) is such that tangentially flowing argon gas from the two anodes forms an inverted 'V-shaped' plasma. The cathode is located directly above the central apex.

DQ 4.3

Do you think that the position of the cathode directly above the two anodes offers any advantage/disadvantage for the DCP?

Answer

The sample aerosol is directed into the apex of the V but does not penetrate the plasma. This can lead to spectrochemical interferences as the sample aerosol does not experience the same temperature environment as in the ICP.

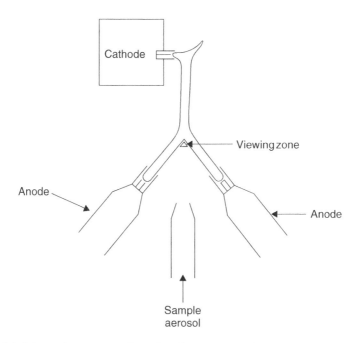

Figure 4.4 Schematic representation of a direct-current plasma [1]. From Dean, J. R., *Atomic Absorption and Plasma Spectroscopy*, 2nd Edition, ACOL Series, Wiley, Chichester, UK, 1997. © University of Greenwich, and reproduced by permission of the University of Greenwich.

4.4 Microwave-Induced Plasma

The microwave-induced plasma (MIP) is usually generated in a disc-shaped cavity resonator, e.g. a 'Beenakker-type' resonator. Located through the cavity is a capillary tube, of approximately 1–2 mm internal diameter, in which the plasma will be formed. Carrier gas is introduced through the capillary tube prior to initiation of the plasma. Typical carrier gases for the MIP include argon and nitrogen, but more probably helium. Power to the cavity is supplied by a microwave generator which supplies 50–200 W of power at 2.45 GHz. Initiation of the plasma, as for the ICP, requires the addition of a spark from a Tesla coil.

The MIP is characterized by a high excitation temperature (7000–9000 K) but a low gas temperature (1000 K). This situation favours the emission of 'non-metals'.

The analysis of non-metals by the MIP has been successfully exploited, such that a coupled gas chromatography–MIP–atomic emission spectroscopy (AES) system is now commercially available (Figure 4.5).

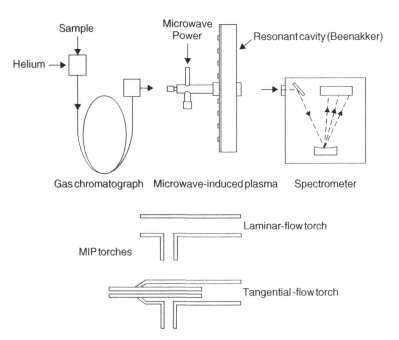

Figure 4.5 Schematic diagram of a coupled gas chromatography–microwave-induced plasma (MIP)–atomic emission spectroscopy system, plus examples of typical MIP torches [1]. From Dean, J. R., *Atomic Absorption and Plasma Spectroscopy*, 2nd Edition, ACOL Series, Wiley, Chichester, UK, 1997. © University of Greenwich, and reproduced by permission of the University of Greenwich.

4.5 Glow Discharge

Glow discharges are essentially low-pressure plasmas that rely on cathodic 'sputtering' to atomize solid samples. The atomized sample can then undergo collision with electrons and metastable fill-gas atoms, so causing excitation and ionization. A schematic diagram explaining the operation of a glow-discharge source is shown in Figure 4.6. Flow of electric current through the glow discharge initiates the analytical process. As the negatively charged electrons are attracted towards the anode (positive terminal), collisions with the (argon) fill-gas atoms can occur, so producing argon ions. The initiation of this plasma provides an integral site for further ionization and excitation. Concurrent with plasma formation, the positively charged argon ions are attracted towards the cathode. As the conducting sample and cathode are one and the same, the argon ions are able to collide with the sample surface, thus liberating metal atoms. This process is known as 'sputtering'. The liberated sample metal atoms can then diffuse into the plasma for excitation/ionization (for AES) or ionization (for MS). The two most probable principal ionization mechanisms are as follows:

$$M + e^- \longrightarrow M^+ + 2e^- \quad \text{(electron impact)} \tag{4.1}$$

or:

$$M + Ar^* \longrightarrow M^+ + Ar + e^- \quad \text{('Penning exchange')} \tag{4.2}$$

where M represents a sputtered atom and Ar^* represents an excited (metastable)-state argon atom.

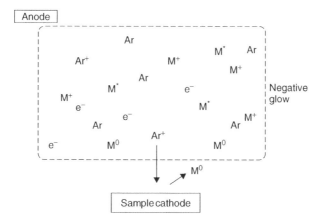

Figure 4.6 Schematic representation of the operating principles of a glow-discharge source [1]. From Dean, J. R., *Atomic Absorption and Plasma Spectroscopy*, 2nd Edition, ACOL Series, Wiley, Chichester, UK, 1997. © University of Greenwich, and reproduced by permission of the University of Greenwich.

Penning ionization is believed to be the dominant exchange process and leads to fairly uniform ionization sensitivity.

Summary

By describing the different approaches that are possible, this chapter highlights the importance of plasma technology for elemental analysis. Although the major focus of this text is the inductively coupled plasma, other plasma sources, namely the direct-current plasma, microwave-induced plasma and glow discharge, are included for completeness. The fundamental aspects of each technique are described.

References

1. Dean, J. R., *Atomic Absorption and Plasma Spectroscopy*, 2nd Edition, ACOL Series, Wiley, Chichester, UK, 1997.
2. Davies, R., Dean, J. R. and Snook, R. D., *Analyst*, **110**, 535–540 (1985).

Chapter 5

Inductively Coupled Plasma–Atomic Emission Spectroscopy

Learning Objectives

- To be able to calculate the energy, wavelength or frequency of a spectral line in appropriate units.
- To be able to define the terms 'absorption' and 'emission' (of radiation).
- To be able to explain and understand the importance of identifying a resonance line (wavelength).
- To be able to calculate the relative populations of the ground and excited states for a spectral line.
- To be able to define spectral line width.
- To appreciate the main types of line broadening in atomic spectroscopy.
- To be able to calculate Doppler and natural line widths.
- To describe the two different approaches in which an ICP can be viewed.
- To be able to calculate the limit of detection and background equivalent concentration in relation to ICP analysis.
- To describe the merits of sequential and simultaneous multi-element detection.
- To be able to recognize the different optical arrangements of spectrometers used for atomic emission spectroscopy.
- To be able to calculate diffraction grating parameters, e.g. resolution.
- To understand the benefits of using an Echelle spectrometer.

Practical Inductively Coupled Plasma Spectroscopy J. R. Dean
© 2005 John Wiley & Sons, Ltd

- To be able to describe the operation of a photomultiplier tube.
- To understand the importance of charge-transfer devices in atomic emission spectroscopy.
- To be able to identify interferences and their remedies in atomic emission spectroscopy.

5.1 Fundamentals of Spectroscopy

The most common emission source for spectroscopic measurement is the inductively coupled plasma (see Section 4.2). The emission arises from specific energy changes within an atomic system. The regions of the electromagnetic spectrum can be identified in terms of a wavelength and a frequency (Figure 5.1). A relationship exists that allows wavelength (λ) and frequency (f) to be determined, provided that one of the terms is known. Wavelength is normally expressed in units of metres (m) and frequency in cycles per second (s^{-1}) or hertz (Hz). The relationship is as follows:

$$c = f\lambda \tag{5.1}$$

where c, the velocity of light, approximates to $3.00 \times 10^8 \, \text{m s}^{-1}$.

SAQ 5.1

If the frequency of electromagnetic radiation is 5×10^{14} Hz, what is the wavelength of this radiation and in which spectral region does it occur?

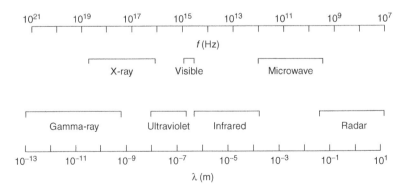

Figure 5.1 Regions of the electromagnetic spectrum [1]. From Dean, J. R., *Atomic Absorption and Plasma Spectroscopy*, 2nd Edition, ACOL Series, Wiley, Chichester, UK, 1997. © University of Greenwich, and reproduced by permission of the University of Greenwich.

As well as frequency and wavelength, electromagnetic radiation can also be expressed in terms of 'packets' of energy (E) called *photons* (or *quanta*). The energy of a photon can be expressed in terms of frequency, as follows:

$$E = hf \qquad (5.2)$$

where h is the Planck constant (6.626×10^{-34} J s).

By substitution Equation (5.1) into Equation (5.2) we can obtain an expression related directly to the wavelength:

$$E = hc/\lambda \qquad (5.3)$$

5.1.1 Origins of Atomic Spectra

If an atom is supplied with sufficient (thermal) energy, the electron is raised from a low-energy level (e.g. ground state) to one with a higher energy (excited state). This is referred to as *absorption*. As the excited state is unstable, the electron returns to a lower-energy state (by inference, a more stable situation). This is referred to as *emission*. Both absorption and emission occur at certain selected wavelengths, frequencies or energies (Figure 5.2).

At room temperature, all of the atoms of a sample are in the ground state. For example, the single outer electron of sodium occupies the 3s orbital (*note*: the electron configuration for Na is $1s^2\ 2s^2\ 2p^6\ 3s^1$). In a hot environment (e.g. an ICP), the sodium atoms are capable of *absorbing* radiation, such that electronic transitions from the 3s level to higher excited states can occur. These electronic transitions occur at specific wavelengths. Experimental observation of sodium

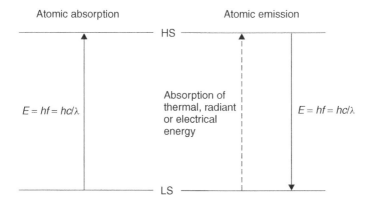

Figure 5.2 Schematic representation of atomic absorption and atomic emission energy transitions: LS, lower-energy state or ground state; HS, higher-energy state [1]. From Dean, J. R., *Atomic Absorption and Plasma Spectroscopy*, 2nd Edition, ACOL Series, Wiley, Chichester, UK, 1997. © University of Greenwich, and reproduced by permission of the University of Greenwich.

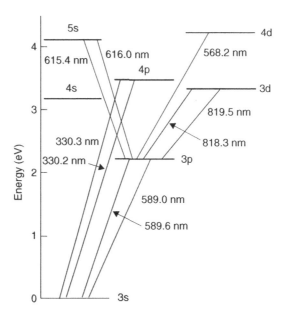

Figure 5.3 Energy-level diagram for sodium [1]. From Dean, J. R., *Atomic Absorption and Plasma Spectroscopy*, 2nd Edition, ACOL Series, Wiley, Chichester, UK, 1997. © University of Greenwich, and reproduced by permission of the University of Greenwich.

identifies absorption peaks at 589.0, 589.6, 330.2 and 330.3 nm. By considering the energy level diagram in Figure 5.3 for sodium, it is possible to identify that these wavelength *doublets* correspond to electronic transitions from the 3s level to either the 3p or 4p levels for 589.0/589.6 nm and 330.2/330.3 nm, respectively. While other electronic transitions are possible, the strongest, i.e. most intense, absorption spectrum occurs for electronic transitions from the ground state (3s) to upper levels. The wavelengths at which these transitions occur are called *resonance lines*.

In the hot environment of an ICP, the electron is easily excited to an upper energy level. However, as the lifetime of the excited atom is brief (typically 10^{-8} s) its return to the ground state is accompanied by the emission of a photon of radiation. In Figure 5.4, the wavelength doublet at 590 nm (589.0 and 589.6 nm in Figure 5.3) represents the most intense emission lines for sodium and is responsible for the yellow colour when sodium salts are introduced into a flame, e.g. in a Bunsen burner.

SAQ 5.2

Confirm that the energy difference between the 3p and 3s levels in Figure 5.4 corresponds to the expected wavelength (*note*: 1 eV $= 1.602 \times 10^{-19}$ J).

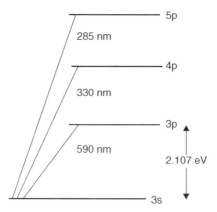

Figure 5.4 Simplified energy-level diagram for sodium [1]. From Dean, J. R., *Atomic Absorption and Plasma Spectroscopy*, 2nd Edition, ACOL Series, Wiley, Chichester, UK, 1997. © University of Greenwich, and reproduced by permission of the University of Greenwich.

It is important to note that both the emission and absorption lines for sodium occur at identical wavelengths since the transitions involved are between the same energy levels.

Emission spectra, in particular, can be further complicated by the presence of both band and continuous spectra. Band (or molecular) spectra arise from the excitation of molecular species in the hot environment of the source. Thus, it is common to observe band spectra from diatomic molecules. In order to differentiate these molecular species from the atomic species of interest, it is necessary to use a high-resolution spectrometer (see Section 5.3). Typical molecular species encountered are C_2 molecules (if organic solvents are introduced) and OH radicals. Their appearance can be troublesome, producing undesirable interference effects. Spectra continua which can arise from recombinations, such as electrons and ions that form atoms and Bremsstrahlung† in plasmas, generally cause elevation of the background against which the emission lines are measured.

5.1.2 Spectral Line Intensity

Spectral line intensities depend on the relative populations of the ground or lower electronic state and the upper excited state. The relative populations of the atoms in the ground or excited states can be expressed in terms of the Boltzmann distribution law, as follows:

$$N_1/N_0 = g_1/g_0 e^{(-\Delta E/kT)} \tag{5.4}$$

---

† Continuous background emission – electromagnetic radiation arising from collision or deviation between fast-moving electrons or atoms (from the German: 'braking radiation').

where N_1 is the number of atoms in the excited state, N_0 the number of atoms in the ground or lower state, g_1 and g_0 are the number of energy levels having the same energies for the upper (excited)- and lower (ground)-energy levels, respectively (*note*: energy levels of the same energy are usually referred to as being *degenerate*), ΔE the difference in energy between the lower-and upper-energy states, k the Boltzmann constant $(8.314\,\mathrm{J\,K^{-1}\,mol^{-1}})$, and T the temperature.

Using sodium as an example, there are two degenerate 3p energy levels (excited state) and a single ground state, 3s, producing a g_1 value of 2 and a g_0 value of 1. Simplification of the above equation gives the following:

$$N_1/N_0 = 2\ e^{(-\Delta E/kT)} \tag{5.5}$$

SAQ 5.3

Calculate the ratio of the populated upper and ground transition states for the spectral transition at 589 nm at a typical plasma temperature of 7000 K.

5.1.3 Spectral Line Broadening

From the discussion so far, you might have the impression that emission line profiles are very narrow and occur at particular discrete wavelengths. This is in essence true, but due to other processes that occur the observed spectral lines profiles are invariably broadened. Figure 5.5 shows the typical shape of a spectral line. The line width is defined as 'the width at half the peak height' ($\Delta\lambda_{1/2}$). Three main factors influence the line widths, namely natural, Doppler and pressure broadening.

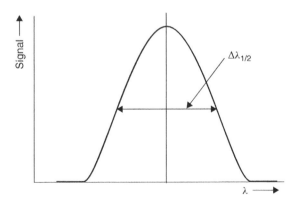

Figure 5.5 Typical shape of a spectral line [1]. From Dean, J. R., *Atomic Absorption and Plasma Spectroscopy*, 2nd Edition, ACOL Series, Wiley, Chichester, UK, 1997. © University of Greenwich, and reproduced by permission of the University of Greenwich.

The *natural* line width broadening is a consequence of the short lifetime (approximately 10^{-8} s) of an atom in an excited state. In 1927, Werner Heisenberg postulated that *nature places limits on the precision with which certain pairs of physical measurements can be made*. Due to this *Uncertainty Principle*, the natural width of an emission line ($\Delta\lambda_N$) can be determined from the following expression:

$$\Delta\lambda_N = \lambda^2 \Delta v / c \qquad (5.6)$$

where λ is the wavelength of the emission line, Δv the uncertainty in the frequency of the emitted radiation (*note*: Δv is equal to $1/\tau$, where τ is the lifetime of the excited state), and c the speed of light.

SAQ 5.4

What is the natural line width for a sodium emission line of wavelength 589 nm and an excited lifetime of 2.5×10^{-9} s?

It should be noted that the observed line width for sodium, at 589 nm, is some 10 times wider than the value found from SAQ 5.4 (assuming that you have carried out the calculation correctly!). Therefore, other broadening processes must predominate.

The thermal motion of atoms in a gas (or plasma) introduces an additional broadening of the line profile – *Doppler* broadening ($\Delta\lambda_D$). An equation that describes this broadening is given below, in units of metres:

$$\Delta\lambda_D = (2\lambda/c)\sqrt{2RT/M} \qquad (5.7)$$

where λ is the wavelength of the emission or absorption, c the speed of light (3×10^8 m s^{-1}), R the gas constant (8.314 J K^{-1} mol^{-1}), T the temperature and M the atomic mass of an atom.

SAQ 5.5

Calculate the Doppler line width for the 589 nm spectral line of the sodium atom in a plasma of temperature 7000 K (*note*: the atomic mass of sodium is 23 g mol^{-1}, but in SI units the value is 23×10^{-3} kg mol^{-1}).

If you have carried out the above calculation correctly, it should now be evident that the major contributor to the observed line width (0.005 nm) for sodium in a plasma is Doppler broadening.

Pressure broadening arises from collisions between the emitting species with other atoms or ions in the plasma. These collisions cause small changes in the ground-state energy levels and hence a subsequent small variation in the emitted wavelength. In plasmas, the collisions are between the atoms of interest and the argon of the plasma (*note*: this type of broadening is usually referred to as *Lorentz* broadening). This results in significant broadening (similar, but slightly less than that obtained for Doppler broadening – typically 3 pm for sodium).

5.2 Plasma Spectroscopy

The concept of using an inductively coupled plasma as a source for atomic emission spectroscopy is attributable to the independent work of Greenfield *et al.* [2] in the UK and Wendt and Fassel [3] in the USA. However, they were not the first to use an ICP torch – this was first reported by Reed [4] who used the technique for growing crystals under high-temperature conditions. While these early instruments would probably not be recognizable in the modern laboratory, the concept of ICP–AES as an elemental technique was born more than 40 years ago.

Light emitted from the plasma source is focused onto the entrance slit of a spectrometer by using a convex lens arrangement. However, two viewing modes are possible (see also Section 4.2). In 'side-on' or 'lateral' viewing, light from the plasma source is orthogonal to the central channel of the ICP, whereas in 'axial' or 'end-on' viewing light from the plasma source is coincident with the central channel of the ICP (Figure 5.6). While both possibilities exist, it is the side-on approach that is most commonly used. Figure 4.3 compares the typical background spectra obtained by side-on and end-on viewing. An evaluation of radially and axially viewed ICPs has recently been reported [5]. Table 5.1 shows the experimental configurations for two plasma-based systems using axially and radially viewed configurations. Diagnostic tests were performed and these are summarized in Table 5.2. The *limit of detection* (LOD) and *background equivalent concentration* (BEC) were calculated by using the following definitions, respectively:

$$BEC = C_{rs}/SBR \tag{5.8}$$

$$LOD = (3 \times BEC \times RSD)/100 \tag{5.9}$$

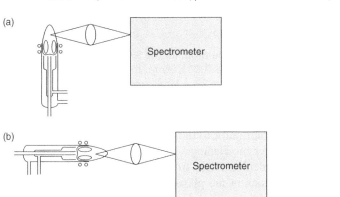

Figure 5.6 Schematic representations of the two viewing modes for inductively coupled plasmas: (a) axially (end-on); (b) radially or lateral (side-on).

Table 5.1 Comparison of the experimental configurations used for axially and radially viewed inductively coupled plasma sources for atomic emission spectroscopy [5]

Optical system	Parameter
Polychromator	Echelle grating plus CaF_2 cross-dispersing prism
Grating density groove	95 grooves mm^{-1}
Focal length	400 mm
Entrance slit	Height, 0.029 mm; width, 0.051 mm
Detector	Peltier cooled charge-coupled device; 70 908 pixels spread across 70 non-linear arrays; wavelength range, 167–785 nm.

Sample introduction system	Type
Nebulizer	Concentric
Spray chamber	Cyclonic

Plasma conditions	Parameter
Frequency	40 MHz
RF power	1.2 kW
Plasma gas flow rate	15.0 l min^{-1}
Auxiliary gas flow rate	1.5 l min^{-1}
Nebulizer gas flow rate	0.8 l min^{-1}
Sample flow rate	0.8 ml min^{-1}
Torch injector tube diameter	2.3 mm (axially viewed ICP); 1.4 mm (radially viewed ICP)
Observation height[a]	13 mm
Emission lines (examples)[b]	Ar(I) 404.442; Ar(I) 404.597; Ba(II) 230.424; Ba(II) 455.403; Mg(II) 280.264; Mg(I) 285.208; Ni(II) 231.604

[a] Only for the case of a radially viewed ICP.
[b] In nm.

Table 5.2 Diagnostic procedures used for axially and radially viewed inductively coupled plasma sources for atomic emission spectroscopy [5]

'Figure of merit'	Parameter[a,b]
UV spectral resolution	Profile of Ba(II) 230 nm line
Visible spectral resolution	Profile of Ba(II) 455 nm line
Robustness	Mg(II) 280/Mg(I) 285 nm ratio
Short-term stability	RSD for Mg(I) 285 nm emission signal ($n = 15$)
Long-term stability	RSD for Mg(I) 285 nm emission signal ($n = 8$; $t = 2$ h)
Sensitivity	LOD for Ni(II) 231 nm line
'Warm-up' time	RSDs for Ar, Ba and Mg emission lines

[a] RSD, relative standard deviation.
[b] LOD, limit of detection.

where C_{rs} is the concentration of a multi-element reference solution $(20 \, mg \, l^{-1})$, SBR the signal-to-background ratio $(= (l_{rs} - l_{blank})/l_{blank})$, where l_{rs} and l_{blank} are the emission intensities for the multi-element and reference solutions, respectively) and RSD the relative standard deviation for ten measurements of the blank (reference) solution.

The results obtained from the diagnostic tests are shown in Table 5.3. Robustness was quantified by measuring the Mg(II) and Mg(I) emission signals at a radiofrequency (RF) power (to the ICP torch) of 1.2 kW. The nebulizer gas flow rate was then adjusted until the maximum Mg(II)/Mg(I) ratio was achieved (0.90 and 0.70 l min^{-1} for the axially and radially viewed ICP systems, respectively). The resultant Mg ratio was then multiplied by 1.8 to correct for response intensities resulting from the use of an Echelle spectrometer with a charge-coupled device (CCD) detector [6]. It can be concluded that both of the axially and radially viewed plasma configurations offer similar figures of merit. Some improvements in sensitivity are noted (Table 5.3 and 5.4) for the axially viewed plasma. However, the radially viewed plasma is ready for use in a shorter time ('warm-up' time) than the axially viewed plasma. Despite these differences, the

Table 5.3 Results obtained from the diagnostic tests camed out on axially and radially viewed inductively coupled plasma sources for atomic emission spectroscopy [5]

'Figure of merit'	Axially viewed[a]	Radially viewed[a]
UV spectral resolution	8 pm	8 pm
Visible spectral resolution	30 pm	30 pm
Robustness[b]	10.6	13.7
Short-term stability	0.70% RSD	0.60% RSD
Long-term stability	1.5% RSD	1.4% RSD
Sensitivity[c]	$0.23 \, \mu g \, l^{-1}$	$3.8 \, \mu g \, l^{-1}$
'Warm-up' time[d]	20 min	10 min

[a] RSD, relative standard deviation.
[b] Multiplied by 1.8.
[c] See also Table 5.4.
[d] Emission intensity deviations less than 5%.

Table 5.4 Limit of detection (LOD) and background equivalent concentration (BEC) data for Ni(II) (231.604 nm) in 0.14 mol l^{-1} of HNO$_3$ and 1000 mg l^{-1} of Cr media for axially and radially viewed inductively coupled plasma sources for atomic emission spectroscopy [5]

Media	BEC ($\mu g \, l^{-1}$)		LOD ($\mu g \, l^{-1}$)	
	Axially	Radially	Axially	Radially
HNO$_3$	11	301	0.23	3.8
Cr	4	741	0.31	7.6

Table 5.5 Analysis of two certified (standard) reference materials using axially and radially viewed inductively coupled plasma–atomic emission spectroscopy [5]

Element (concentration)	Axially viewed	Radially viewed	Certificate value
NIST 1515 Apple Leaves[a]			
Calcium (wt%)	1.32 ± 0.06	1.46 ± 0.03	1.526 ± 0.015
Copper (mg kg^{-1})	4.98 ± 0.23	6.14 ± 0.06	5.64 ± 0.24
Iron (mg kg^{-1})	61.9 ± 1.4	66.5 ± 4.3	83 ± 5
Magnesium (wt%)	0.241 ± 0.010	0.247 ± 0.003	0.271 ± 0.008
Manganese (mg kg^{-1})	44.5 ± 1.1	48.5 ± 2.5	54 ± 3
Zinc (mg kg^{-1})	10.9 ± 0.3	19.6 ± 2.7	12.5 ± 0.3
NIST 1577b Bovine liver[a]			
Calcium (mg kg^{-1})	146 ± 12	107 ± 5	116 ± 4
Copper (mg kg^{-1})	142 ± 1	148 ± 1	160 ± 8
Iron (mg kg^{-1})	164 ± 4	156 ± 4	184 ± 15
Magnesium (mg kg^{-1})	546 ± 5	523 ± 4	601 ± 28
Manganese (mg kg^{-1})	7.20 ± 0.58	9.14 ± 0.75	10.5 ± 0.17
Zinc (mg kg^{-1})	114 ± 1	110 ± 3	127 ± 16

[a]NIST, National Institute of Science and Technology.

analytical performance of each system was not compromised. Both ICP systems were then applied to the analysis of two certified reference materials which had previously been microwave-digested in a closed-vessel system using nitric acid and hydrogen peroxide, to assess accuracy and precision. The results, shown in Table 5.5, indicate that both systems were capable of providing accurate and precise results.

5.3 Spectrometers

The spectrometer is required to separate the emitted light into its component wavelengths. In practice, two options are available. The first of these involves a capability to measure one wavelength, corresponding to one element at a time, while the second allows multi-wavelength or multi-element detection. The former is known as *sequential analysis* or *sequential multi-element analysis* if the system is to be used to measure several wavelengths one at a time, while the latter is termed *simultaneous multi-element analysis*. The typical wavelength coverage of a spectrometer for atomic emission spectroscopy (AES) is between 167 nm (Al) to 852 nm (Cs).

SAQ 5.6

Do you envisage any difficulties in operating a spectrometer below 190 nm?

Separation of the light into its component wavelengths is achieved in all modern instruments by the use of a diffraction grating. The latter consists of a series of closely spaced lines ruled or etched onto the surface of a mirror. Most gratings for AES have a line, or groove, density between 600 and 3200 lines mm^{-1}. When light strikes the grating, it is diffracted at an angle which is dependent on the grating equation, as follows:

$$n\lambda = d \sin \phi \qquad (5.10)$$

where n is the spectral order, λ the wavelength, d the distance between a line or groove on the grating and ϕ the angle of a groove.

By using Equation (5.10), it is possible to calculate the expected wavelength of a spectral line.

SAQ 5.7

Calculate the wavelength of a spectral emission line, in the first order, with a groove density of 1200 lines mm^{-1} and an angle, ϕ, of 20°.

Interference or 'ghost' images, from overlapping wavelengths can occur. To prevent spectral overlap, it is possible to use blazed reflection gratings. In this situation, the grooves are ruled at a specified angle (known as the Blaze angle) and appear as a 'saw-tooth' pattern. Thus, it is possible to have a blazed diffraction grating which is more efficient in a specific wavelength region. The typical arrangement of the diffraction grating is shown in Figure 5.7 – the exception to this is the Echelle grating which is described in Section 5.3.2. The resolution of

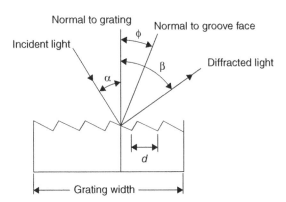

Figure 5.7 Schematic diagram of a blazed grating: d, distance between grooves; ϕ, angle of a groove (blaze angle); α, angle of incidence; β, angle of reflection [1]. From Dean, J. R., *Atomic Absorption and Plasma Spectroscopy*, 2nd Edition, ACOL Series, Wiley, Chichester, UK, 1997. © University of Greenwich, and reproduced by permission of the University of Greenwich.

the grating is related to the spectral order (m) and the total number of grooves (N), as follows:

$$R = mN \qquad (5.11)$$

while the resolving power, $\Delta\lambda$, is defined as the wavelength divided by the resolution.

SAQ 5.8

Calculate (a) the resolution for a conventional grating ruled with 1200 lines mm^{-1} with a width of 52 mm in the first order, and (b) the resolving power at 300 nm.

5.3.1 Sequential

A *sequential* spectrometer is the lower-cost option for AES. This typically consists of entrance and exit optics, a diffraction grating and a single detector. A sequential spectrometer has the advantage of flexibility in terms of wavelength coverage. Selection of the desired wavelength is achieved by rotation of the grating within its spectrometer mounting. This rotation can be achieved manually, or more typically in modern instruments, by computer control.

DQ 5.1

Can you forsee any potential disadvantages of operating with a sequential spectrometer.

Answer

In order to improve the precision of the data it is often necessary to include an internal standard (an element that is not present in the sample); other recorded signals can then be referenced to this standard. This reduces any potential interferences from, for example, a change in viscosity from sample to sample. However, in sequential analysis only one element at a time can be monitored, and therefore the use of an internal standard is not possible. (In practice, however, software control will allow two wavelengths to be monitored, thus allowing 'pseudo-wavelength' information to be obtained.)

The Czerny–Turner configuration is the most common spectral mounting for sequential AES. The optical layout of a spectrometer which incorporates these features is shown in Figure 5.8.

5.3.2 Simultaneous

One of the major advantages of atomic emission spectroscopy is the ability to perform *simultaneous* multi-element analysis. In simultaneous analysis, many

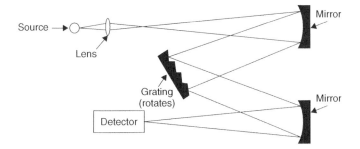

Figure 5.8 Schematic diagram of the optical layout of a spectrometer which incorporates the Czerny–Turner spectral mounting (configuration) [1]. From Dean, J. R., *Atomic Absorption and Plasma Spectroscopy*, 2nd Edition, ACOL Series, Wiley, Chichester, UK, 1997. © University of Greenwich, and reproduced by permission of the University of Greenwich.

wavelengths or elements (typically, 20–70) can be monitored at the same time. More than one wavelength may be specified for each element if, for example, spectral interference from another element is known to occur. The limitation of such a system is that the exit slits are pre-set and this allows no flexibility if another element and/or wavelength is required to be analysed. This approach has traditionally been carried out by using a polychromator. The Paschen–Runge mounting is the most commonly used polychromator. The grating, entrance slit and multiple exits slits are fixed around the periphery of what is known as a 'Rowland circle'. The grating is concave in appearance and does not rotate. The optical layout of a spectrometer which incorporates these features is shown in Figure 5.9.

An alternative approach for simultaneous multi-element analysis is the Echelle spectrometer. While this spectrometer was originally only commercially available for the direct-current plasma (DCP), it has become increasingly important because of the special spectral features that are inherent in its design. The major component difference of the echelle spectrometer is the grating. Unlike the previous design outlined above, this grating utilizes the spectral order (recall the grating equation (Equation (5.10)) for maximum wavelength coverage. A typical Echelle grating is ruled with only 50–100 lines or grooves per mm. The resolution of a diffraction grating is directly related to the groove density (n) and the spectral order (M) (Equation (5.11)). In this situation, however, instead of using a grating with a large number of grooves, the resolution is improved by increasing the blaze angle and spectral order. The arrangement of an Echelle grating is shown in Figure 5.10. If you compare this with the blazed grating shown in Figure 5.7, it is obvious that the light is reflected off the 'short-side' of the grating (and not the 'long-side', as in Figure 5.7). Therefore, the blaze angle is greater than $45°$. The advantages of using this method to improve spectral resolution can be seen in Table 5.6.

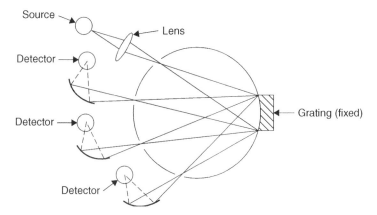

Figure 5.9 Schematic diagram of the optical layout of a spectrometer which incorporates the Paschen–Runge spectral mounting (configuration) [1]. From Dean, J. R., *Atomic Absorption and Plasma Spectroscopy*, 2nd Edition, ACOL Series, Wiley, Chichester, UK, 1997. © University of Greenwich, and reproduced by permission of the University of Greenwich.

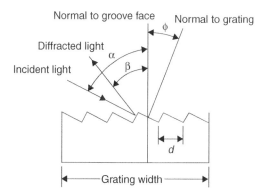

Figure 5.10 Schematic diagram of an Echelle grating: d, distance between grooves; ϕ, angle of a groove (blaze angle); α, angle of incidence; β, angle of reflection [1]. From Dean, J. R., *Atomic Absorption and Plasma Spectroscopy*, 2nd Edition, ACOL Series, Wiley, Chichester, UK, 1997. © University of Greenwich, and reproduced by permission of the University of Greenwich.

However, in order to prevent overlapping of spectral orders a secondary dispersion is required. This is typically carried out by using a prism. If the latter is placed so that its separation occurs perpendicular to the diffraction grating, a two-dimensional spectral 'map' is produced (Figure 5.11). The spectral 'map' generated is thus sorted into spectral order vertically and wavelength horizontally (Figure 5.12).

Table 5.6 Comparison of the spectral features of a conventional diffraction grating and an Echelle grating [1]. From Dean, J. R., *Atomic Absorption and Plasma Spectroscopy*, 2nd Edition, ACOL Series, Wiley, Chichester, UK, 1997. © University of Greenwich, and reproduced by permission of the university of Greenwich

Parameter	Conventional grating	Echelle grating
Focal length (m)	0.5	0.5
Groove or line density (lines mm^{-1})	1200	79
Diffraction angle	10° 22′	63° 26′
Width (mm)	52	128
Spectral order[a]	1	75
Resolution	62 400	758 400
Resolving power (nm)[a]	0.004 81	0.000 396

[a] At 300 nm.

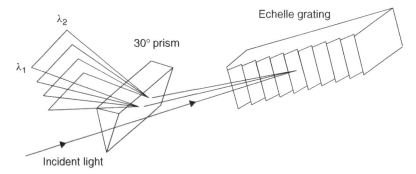

Figure 5.11 Schematic representation of the two-dimensional dispersion obtained by using a prism in conjunction with an Echelle grating [1]. From Dean, J. R., *Atomic Absorption and Plasma Spectroscopy*, 2nd Edition, ACOL Series, Wiley, Chichester, UK, 1997. © University of Greenwich, and reproduced by permission of the University of Greenwich.

DQ 5.2

Do you know which type of detection system might allow you to observe such a two-dimensional spectral 'map'?

Answer

Charge-transfer devices can be used to monitor the detailed information required to generate such a 'map'; these will be described in the next section.

A typical optical layout of the Echelle spectrometer is shown in Figure 5.13.

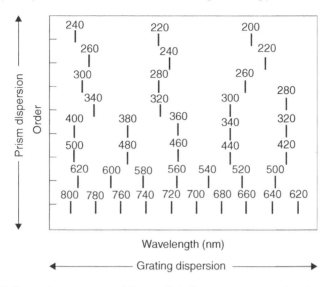

Figure 5.12 Spectral map generated by the Echelle spectrometer, using the arrangement shown in Figure 5.11 [1]. From Dean, J. R., *Atomic Absorption and Plasma Spectroscopy*, 2nd Edition, ACOL Series, Wiley, Chichester, UK, 1997. © University of Greenwich, and reproduced by permission of the University of Greenwich.

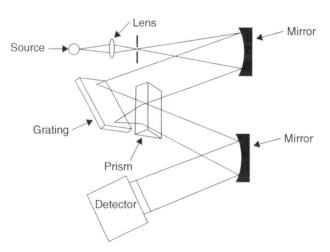

Figure 5.13 Schematic diagram of the optical layout of the Echelle spectrometer [1]. From Dean, J. R., *Atomic Absorption and Plasma Spectroscopy*, 2nd Edition, ACOL Series, Wiley, Chichester, UK, 1997. © University of Greenwich, and reproduced by permission of the University of Greenwich.

5.4 Detectors

After wavelength separation has been achieved, it is necessary to 'view' the spectral information. Two different types of detectors are used in AES. Up until the 1990s, the most common detector was the photomultiplier tube (PMT); however, more recently this has largely been replaced by multi-channel detectors based on charge-transfer device (CTD) technology. Both a charge-coupled device (CCD) and a charge-injection device (CID) are used for multi-channel detection in AES.

For the different spectrometer configurations described in the previous section, either type of detector can be used. In each case, the detector (PMT or CTD) is mounted behind the exit slit of the spectrometer. Thus, for the Czerny–Turner mounting (Figure 5.8) a single detector is required, while for the Paschen–Runge mounting (Figure 5.9) with 30 exits slits, 30 detectors are required. The latter case obviously adds to the cost and complexity of a polychromator system. However, the major advantage of CTD technology, i.e. multi-wavelength detection with a single detector, has most effectively been exploited by using an Echelle spectrometer. The capability of an Echelle spectrometer to generate a two-dimensional spectral 'map', coupled with a sensitive multi-wavelength detector (i.e. a CCD or CID), provides a complete fingerprint of a sample.

5.4.1 Photomultiplier Tube

The photomultiplier tube (PMT) is a device that converts incident light into a current. This is achieved by a series of steps that can be outlined by consideration of Figure 5.14. Incident light passes through the silica window and strikes the

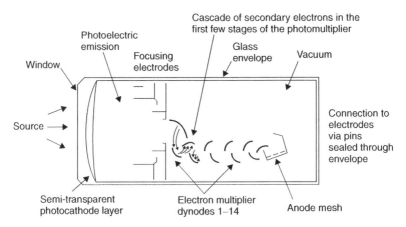

Figure 5.14 Schematic representation of the operation of a photomultiplier tube [1]. From Dean, J. R., *Atomic Absorption and Plasma Spectroscopy*, 2nd Edition, ACOL Series, Wiley, Chichester, UK, 1997. © University of Greenwich, and reproduced by permission of the University of Greenwich.

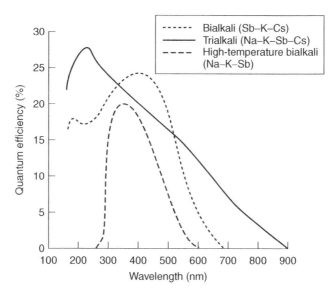

Figure 5.15 Spectral responses of selected photocathode materials [1]. From Dean, J. R., *Atomic Absorption and Plasma Spectroscopy*, 2nd Edition, ACOL Series, Wiley, Chichester, UK, 1997. © University of Greenwich, and reproduced by permission of the University of Greenwich.

photocathode. The latter consists of an easily ionized material such as an alloy of two (or three) alkali metals with antimony. The exact composition of the photocathode affects the wavelength coverage of any particular detector. For example, in Figure 5.15 it can be seen that the use of the bialkali Sb–K–Cs provides a greater wavelength coverage than does the use of the *high-temperature* bialkali Na–K–Sb. In this manner, it is possible to select a PMT that has optimum response characteristics for a particular wavelength range. This feature can easily be exploited in a polychromator where an individual PMT is used to monitor only one wavelength.

SAQ 5.9

After considering the spectral information shown in Figure 5.15, which do you think would be the best photomultiplier tube for use in a monochromator?

A suitably energetic incident photon of light causes an electron to be emitted from the cathode surface. This process is known as the *photoelectric effect*, while the yield per incident photon is called the *quantum efficiency*. A typical value for the latter is between 10 and 25% at 650–340 nm. By a series of focusing electrodes, the emitted electron is directed towards the dynode chain, where the latter acts to multiply this single electron into many electrons. This is achieved because a single electron striking dynode 1 will emit at least two secondary electrons. These electrons will then strike dynode 2 and for each electron that strikes the second dynode, at least a further two electrons are emitted, and so on. In this manner, a cascade of electrons are produced. The exact number of electrons depends on the length of the dynode chain, which typically consists of 9–16 dynodes. This amplification of electrons by the dynode chain is known as the 'gain'. A typical gain for a nine-dynode PMT is 10^6. The final step is to collect the electrons at the anode. The electrical current that is measured at the anode is proportional to the amount of light that struck the photocathode. This current is then converted into a voltage signal which is then transformed via an analogue-to-digital (A/D) converter to a suitable computer for processing purposes.

5.4.2 Charge-Transfer Devices

The utilization of so-called charge transfer devices (CTDs) is perhaps the most significant advance in detector technology for AES. Charge-transfer devices offer a high sensitivity and a wide wavelength coverage, i.e. UV to visible spectra. Their main application has been as detectors for the Echelle spectrometer, although they have been used on polychromator systems using linear charge-coupled device (CCD) arrays around the Rowland circle (see above). The compact nature of the spectral 'map' generated by the Echelle spectrometer can be focused onto a single CTD. All CTDs are semiconductor devices consisting of a series of cells or pixels which accumulate charge when exposed to light. The amount of stored (accumulated) charge is then a measure of the amount of light to which a particular pixel has been exposed. A CTD is an array of closely spaced metal–insulator–semiconductor diodes formed on a wafer of semiconductor material. In operation, a CTD must be exposed to light for a specific period of time, and then 'read'. During this 'read' time, the detector is not exposed to the incoming light. Two common forms are available, i.e. the charge-coupled device (CCD) and the charge-injection device (CID). In the case of a CCD, light from the plasma source is gathered for a specified time and then 'read out' on a 'row-by-row' basis, so allowing the accumulated charge to be transferred from pixel to pixel until it reaches the 'read-out' amplifier. A CID is slightly different in its operation in that it has additional circuitry which allows for the 'read-out' of individual pixels, i.e. the CID pixels can be interrogated individually at any time during exposure to the plasma source.

The major advantages of charge-transfer devices are as follows:

- Flexibility in analytical line (wavelength) selection.

- Use of several lines (wavelengths) for the same element in order to extend the linear dynamic range.

- Use of several lines (wavelengths) for the same element to improve accuracy and to identify potential matrix or spectral interferences.

- Potential for qualitative analysis.

- Plasma-based diagnostic studies, e.g. temperature measurement (see Section 4.1).

However, CTDs also suffer from some limitations relating to their function and operation [7].

5.5 Interferences

Spectral interferences for atomic emission spectroscopy can be classified into two main categories, i.e. spectral overlap and matrix effects. *Spectral interferences* are probably the most well known and best understood. The usual remedy to alleviate a spectral interference is to either increase the resolution of the spectrometer or to select an alternative spectral emission line. Three types of spectral overlap can be identified, as follows (Figure 5.16):

(i) direct wavelength coincidence from an interfering emission line;

(ii) partial overlapping of the emission line of interest from an interfering line in close proximity;

(iii) the presence of an elevated or depressed background continuum.

Type (i) and (ii) interferences can occur as a result of an interfering emission line from another element, the argon source gas, or impurities within or entrained in the source, e.g. molecular species, such as OH and N_2. Extensive work has meant that wavelength coincidence (type (i)) is well characterized for the ICP, for example, Cd (at 228.802 nm) and As (at 228.812 nm); Zn (at 213.856 nm) and Ni (at 213.858 nm). Elimination of the type (ii) interference is usually only possible by an improvement in resolution. As this may not be possible on a routine basis, mathematical models can be used to try and correct for this type of interference. The only certain remedy, however, is to select an interference-free wavelength for the selected element. The type (iii) interference can be corrected for by measurement of the background on either side of the wavelength of interest. Provided that no significant fine structure is present on the background, this method of correction should prove to be satisfactory.

Matrix interferences are often associated with the sample introduction process. For example, pneumatic nebulization can be affected by the dissolved-solids content of the aqueous sample which affects the uptake rate of the nebulizer, and hence the sensitivity of the determination. Matrix effects that are encountered in the plasma source have also been documented. Typically, this involved the presence of easily ionizable elements (EIEs), e.g. alkali metals, within the plasma source. Some specific work has investigated the effect of EIEs on signal

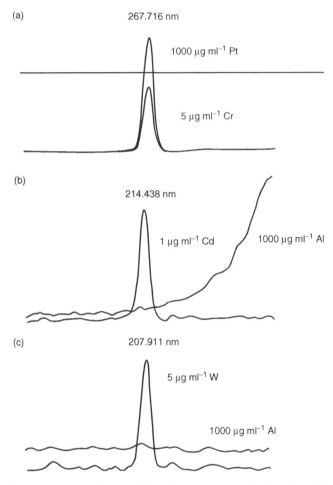

Figure 5.16 Different types of spectral interferences: (a) spectral overlap; (b) wing overlap; (c) background shift [1]. From Dean, J. R., *Atomic Absorption and Plasma Spectroscopy*, 2nd Edition, ACOL Series, Wiley, Chichester, UK, 1997. © University of Greenwich, and reproduced by permission of the University of Greenwich.

suppression or enhancement for both ICP and DCP sources. The effects are greatest in the DCP source where the addition of lithium or barium salts is used as a buffer to reduce the problem of signal enhancement.

Summary

The use of the inductively coupled plasma for atomic emission spectroscopy is highlighted in this chapter, and the fundamentals of spectroscopy relating to its use with an inductively coupled plasma are discussed. Of current interest are the optical viewing positions of the plasma relative to the spectrometer. The main spectrometer designs for inductively coupled plasma–atomic emission spectroscopy are described, with particular emphasis being placed on simultaneous and sequential systems. The use of charge-transfer technology for optical detection in atomic emission spectroscopy is also emphasized.

References

1. Dean, J. R., *Atomic Absorption and Plasma Spectroscopy*, 2nd Edition, ACOL Series, Wiley, Chichester, UK, 1997.
2. Greenfield, S., Jones, L. I. and Berry, C. T., *Analyst*, **89**, 713–720 (1964).
3. Wendt, R. H. and Fassel, V. A., *Anal. Chem.*, **37**, 920–922 (1965).
4. Reed, T. B., *Appl. Phys.*, **32**, 821–824 (1961).
5. Silva, F. V., Trevizan, L. C., Silva, C. S., Nogueira, A. R. A. and Nobrega, J. A., *Spectrochim. Acta*, **57**, 1905–1913 (2002).
6. Dennaud, J., Howes, A., Poussel, E. and Mermet, J. M., *Spectrochim. Acta*, **56**, 101–112 (2001).
7. Mermet, J. M., *J. Anal. At. Spectrom.*, **20**, 11–16 (2005).

Chapter 6

Inductively Coupled Plasma–Mass Spectrometry

Learning Objectives

- To understand the relationship between atomic number, atomic weight and isotopes.
- To appreciate the variation in the 1st and 2nd ionization energies for selected elements.
- To be able to describe the operation of an inductively coupled plasma–mass spectrometer system.
- To understand the requirements of the interface for ICP–MS.
- To understand the principle of operation of a quadrupole mass spectrometer.
- To appreciate the different methods of signal monitoring in mass spectrometry.
- To appreciate the significance of using a sector-field mass spectrometer.
- To be aware of other types of mass spectrometers for ICP–MS.
- To be able to describe the operation of an electron multiplier tube.
- To be able to identify isobaric, molecular and matrix interferences in ICP–MS.
- To appreciate the different approaches for alleviating molecular interferences.
- To be aware of the different reactions that are possible in a collision/reaction cell for removal of spectroscopic interferences in ICP–MS.
- To be able to perform an isotope dilution analysis calculation.
- To be able to interpret ICP–MS spectra.

Practical Inductively Coupled Plasma Spectroscopy J. R. Dean
© 2005 John Wiley & Sons, Ltd

6.1 Fundamentals of Mass Spectrometry

Mass spectrometry is a technique for measuring the molecular weight of elements or compounds. Most applications of mass spectrometry are for organic compounds. However, in this present context we are considering the use of mass spectrometry for elemental analysis.

6.1.1 Atomic Structure

Each element in the Periodic Table is composed of atoms, which themselves are composed of a nucleus (a mixture of protons and neutrons) and electrons. As protons and electrons have charge associated with them (a proton is positively charged and an electron negatively charged) in order that the atom is normally neutral, i.e. has no charge, the atom must consist of equal numbers of protons and electrons. It is possible for the same element, however, to have a different mass. In this situation, we can refer to the isotopes of a particular element. In order for the same element to have a different mass, some additional component must be present.

DQ 6.1

How can the same element have a different mass?

Answer

By having a different number of neutrons. For example, chlorine (Cl) is made up of 17 protons and 17 electrons. However, it can exist as two isotopes, each with a different mass. The lighter isotope contains 18 neutrons and is represented as $^{35}_{17}Cl$, while the heavier isotope is represented as $^{37}_{17}Cl$. The subscript represents the atomic number of chlorine, i.e. the number of protons, while the superscript represents the atomic weight, i.e. the total number of protons and neutrons.

DQ 6.2

How many neutrons are present in the heavier isotope of chlorine?

Answer

This isotope contains 20 neutrons.

Therefore, the two isotopes of chlorine have atomic masses of 35 and 37 amu.† But what of their abundances?

---

† amu, atomic mass unit.

SAQ 6.1

If the lighter isotope of chlorine has an abundance of 75.7% while that of the heavier isotope is 24.2%, what is the resultant overall atomic weight of chlorine?

Ionization is the process whereby an electron can be removed from a neutral atom by applying an external source of energy, e.g. an ICP. The resultant ion (a single positive cation if one electron is removed) has the same atomic mass as the original isotope of the element, as the mass of the electron is negligible (9.110×10^{-31} g when compared to the mass of a proton, 1.673×10^{-27} g, or that of a neutron, 1.675×10^{-27} g). The energy required to remove an electron is called the *ionization energy*. By application of further energy, a second electron can be removed from the resultant ion. Examples of selected element ionization energies are presented in Table 6.1.

DQ 6.3

Comment on the ionization energies given in Table 6.1.

Answer

All first ionization energies are in the approximate range 4–10 eV. All second ionization energies are higher (>14 eV).

6.1.2 Terminology

All mass spectrometers are able to separate ions on the basis of their **mass/charge ratio**, *m/z*, i.e. the atomic mass of the element divided by its charge. It is normal

Table 6.1 Selected ionization energies (in eV) for a range of elements

Element	Symbol	Atomic number	1st ionization energy	2nd ionization energy
Lithium	Li	3	5.392	75.622
Sodium	Na	11	5.139	47.292
Potassium	K	19	4.341	31.811
Rubidium	Rb	37	4.177	27.499
Zinc	Zn	30	9.394	17.960
Cadmium	Cd	48	8.993	16.904
Mercury	Hg	80	10.437	18.752
Silicon	Si	14	8.151	16.339
Germanium	Ge	32	7.899	15.93
Tin	Sn	50	7.344	14.629
Lead	Pb	82	7.416	15.04

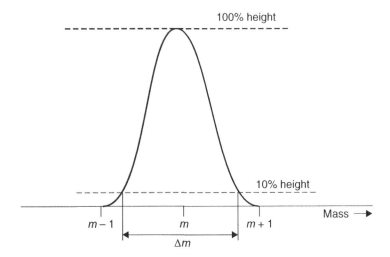

Figure 6.1 Resolution, as defined in mass spectrometry.

that the charge on the element is a single positive one, i.e. a cation; therefore, $z = 1$.

A range of mass spectrometers can be used for inductively coupled plasma–mass spectrometry (ICP–MS) (see Section 6.4). An important characteristic of a mass spectrometer is its ability to separate ions that have similar m/z ratios, i.e. **resolution** (R). This is defined by considering two adjacent m/z ratios, i.e. an isotope with a mass of 'm' and a second isotope (can be the same element or a different one) with a mass of '$m + \Delta m$' (Figure 6.1).

$$R = m / \Delta m \qquad (6.1)$$

6.2 Inorganic Mass Spectrometry

The coupling of a plasma source (capillary arc) to a mass spectrometer was first reported in 1974 [1]. However, Gray and co-workers soon noticed that the use of a capillary arc was a *limited* ion source. The obvious replacement ion source was the inductively coupled plasma (ICP), which at that time had found acceptance in research laboratories as a source for atomic emission spectroscopy. The coupling of an ICP with a mass spectrometer was first reported in 1980, involving collaboration between Gray and the Ames Laboratory, USA [2]. For historical accounts of the development of ICP–MS, the reader is directed towards the work of Gray [3, 4]. Now, some 25 years later, the application of ICP–MS is considered to be routine.

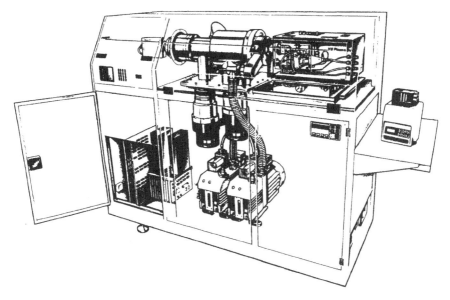

Figure 6.2 A commercially available inductively coupled plasma–mass spectrometer (ICP–MS) system [5]. From 'Applications Literature' published by VG Elemental. Reproduced by permission of Thermo Electron Corporation, Winsford, Cheshire, UK.

In contrast to other sources for inorganic mass spectrometry, the inductively coupled plasma offers several advantages, not least the ability to analyse samples rapidly. This major advantage, coupled with the high degree of sensitivity offered by ICP–MS, are the unique features which have allowed this technique to evolve into the major analytical technique for elemental analysis post-2000.

The major instrumental development required to establish this technique was the efficient coupling of an ICP, operating at atmospheric pressure, with a mass spectrometer, which operates under high vacuum (Figure 6.2). The development of a suitable interface held the key to the establishment of the technique.

An important feature of a mass spectrometer is its ability to measure isotope ratios. The importance of this feature is readily observed when you consider that approximately 70% of the elements in the Periodic Table have stable (non-radioactive) isotopes. The ability to measure isotope ratios has two major benefits, as follows:

- The use of stable isotopes for *tracer* studies, e.g. monitoring the absorption of nutrients in people without the need for radio-labelled isotopes.

- The use of enriched stable isotopes for quantitative analysis, i.e. *isotope dilution analysis* (IDA).

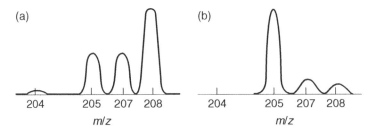

Figure 6.3 Mass spectra of lead isotopes, showing their relative abundances: (a) stable lead; (b) enriched lead.

Figure 6.3 shows the mass spectra of lead isotopes, where Figure 6.3(a) displays the mass spectrum of (stable) lead, together with its relative abundances, while Figure 6.3(b) shows a typical mass spectrum obtained for *enriched* lead, along with its relative abundances. Further details of isotope dilution analysis can be found in Section 6.7.

6.2.1 Ion Source: ICP

The ion source in this present context is the inductively coupled plasma (ICP). Its operation and formation have been described earlier (see Section 4.2). In atomic emission spectroscopy, the ICP can be viewed either axially or radially (see Figure 5.6), while in mass spectrometry the ICP torch is positioned horizontally so that the ions can be extracted from the ICP directly into the mass spectrometer. As a consequence of this horizontal positioning of the ICP torch in relation to the spectrometer, all species that enter the plasma are transferred into the mass spectrometer.

DQ 6.4

Recalling the sample introduction devices used for ICPs, as described earlier, which is the most popular way of achieving this?

Answer

The most common method of introducing a sample into an ICP–MS system is via a pneumatic nebulizer/spray chamber assembly.

6.3 Interface

The main feature in the success of this techniques has been the development of a suitable interface.

DQ 6.5

What would you identify as the main problem in interfacing an inductively coupled plasma with a mass spectrometer?

Answer

The interface allows the coupling of the atmospheric ICP source with the high-vacuum mass spectrometer, while still maintaining a high degree of sensitivity. The interface consists of a water-cooled outer sampling cone which is positioned in close proximity to the plasma source (Figure 6.4). The sampling cone is typically made of nickel because of its high thermal conductivity and relative resistance to corrosion and its robust nature. The pressure differential created by the sampling cone is such that ions from the plasma and the plasma gas itself are drawn into the region of lower pressure through the small orifice of the cone (≈ 1.0 mm). The region behind the sampling cone is maintained at a moderate pressure (≈ 2.5 mbar) by using a rotary vacuum pump. As the gas flow through the sample cone is large, a second cone is placed close enough behind the sampling cone to allow the central portion of the expanding jet of plasma gas and ions to pass through the skimmer cone. The latter (typically made of nickel) has an orifice diameter of ≈ 0.75 mm. The pressure behind the skimmer cone is maintained at $\approx 10^{-4}$ mbar. The extracted ions are then focused by a series of electrostatic lenses into the mass spectrometer. If the ICP–MS system contains a collision/reaction cell (see Section 6.6.3), then this is located before the mass spectrometer.

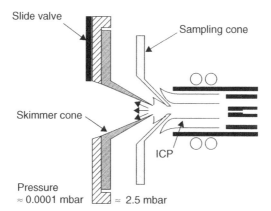

Figure 6.4 Schematic diagram of the inductively coupled plasma–mass spectrometer interface [5]. From Dean, J. R., *Atomic Absorption and Plasma Spectroscopy*, 2nd Edition, ACOL Series, Wiley, Chichester, UK, 1997. © University of Greenwich, and reproduced by permission of the University of Greenwich.

6.4 Mass Spectrometer

The mass spectrometer acts as a filter, transmitting ions with a pre-selected mass/charge ratio. The transmitted ions are then detected and converted into an appropriate form for display.

DQ 6.6

Which type of mass spectrometer would you think could be used for ICP–MS?

Answer

Any type of mass spectrometer can be used. However, the first to be exploited commercially was the quadrupole mass spectrometer. This was closely followed by the high-resolution mass spectrometer in order to overcome some of the deficiencies of a quadrupole mass spectrometer i.e. its inability to overcome interferences (see Section 6.6) due to its low resolution – a quadrupole mass spectrometer being limited to unit mass resolution. More recently, other mass spectrometers have been exploited, namely the ion-trap and time-of-flight mass spectrometers.

6.4.1 Quadrupole Mass Spectrometer

The quadrupole analyser consists of four straight metal rods positioned parallel to and equidistant from the central axis (Figure 6.5). By applying direct current

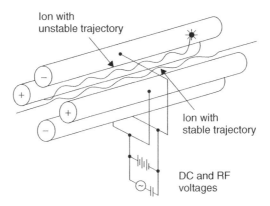

Figure 6.5 Schematic arrangement of the quadrupole analyser arrangement [5]. From Dean, J. R., *Atomic Absorption and Plasma Spectroscopy*, 2nd Edition, ACOL Series, Wiley, Chichester, UK, 1997. © University of Greenwich, and reproduced by permission of the University of Greenwich.

(DC) and radio frequency (RF) voltages to opposite pairs of the rods, it is possible to have a situation where the DC voltage is positive for one pair and negative for the other. Likewise, the RF voltages on each pair are 180° out of phase, i.e. they are opposite in sign, but with the same amplitude. Ions entering the quadrupole are subjected to oscillatory paths by the RF voltage. However, by selecting appropriate RF and DC voltages, only ions of a given mass/charge ratio will be able to traverse the length of the rods and emerge at the other end. Other ions are lost within the quadrupole analyser; as their oscillatory paths are too large, they collide with the rods and become neutralized.

SAQ 6.2

Where do the neutralized ions end up in the system?

ICP–MS can be operated in two distinctly different modes, i.e. with the mass filter transmitting only one mass/charge ratio or with the DC and RF values being changed continuously. The former would allow single-ion monitoring, with the latter allowing multi-element analysis. In single-ion monitoring, all of the data is obtained from a single mass/charge ratio; although this precludes the major facet of the technique, it does provide a higher degree of sensitivity for the element (mass/charge ratio) of interest. For multi-element analysis, RF and DC voltage scanning is required. Scanning, and hence data acquisition, can be carried out in three different modes, as follows:

- a single continuous scan

- peak hopping

- multi-channel scanning

These different scanning modes are illustrated in Figure 6.6, using silver as an example (^{107}Ag is 51.8% abundant and ^{109}Ag 48.2% abundant). In *single continuous scanning*, the mass/charge ratio is changed continuously in one scan. However, in order to reduce ion fluctuations and improve precision it is better if the mass/charge ratio is scanned repetitively. This can be carried out by using either peak hopping or multi-channel scanning. In *peak hopping*, the signal ions are measured at selected mass/charge ratios for a particular 'dwell' time (e.g. 0.5–1 s). This allows fast repetitive analyses of a pre-determined set of elements. However, it does not allow interrogation of the mass spectrum for potential interferences, e.g. unexpected polyatomic interferences (see Section 6.6.2.1). In *multi-channel scanning*, all mass/charge ratios are repetitively scanned, thus providing a complete 'fingerprint' of the unknown sample composition. This mode can also be useful for qualitative analysis of unknown samples. In the multi-channel scanning mode, the typical dwell time per mass/charge ratio is 0.1–0.5 ms. The mass analyser is therefore a very rapid sequential spectrometer.

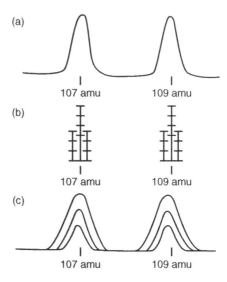

Figure 6.6 Different methods of data acquisition (scanning modes) employed for inductively coupled plasma–mass spectrometry, using silver as an example: (a) single continuous scanning; (b) peak hopping; (c) multi-channel scanning [5]. Reproduced by permission of K. Jarvis from *Handbook of Inductively Coupled Plasma–Mass Spectrometry*, K. E. Jarvis, Gray, A. L. and Houk, R. S., Blackie Academic/Viridian Publishing, p. 45, 1992.

SAQ 6.3

What is the difference between sequential multi-element analysis and simultaneous multi-element analysis? What analogy does this provide in ICP optical spectroscopy?

Quadrupole mass analysers are capable of only unit mass resolution, i.e. they can observe integral values of the mass/charge ratio only (e.g. 204, 205, 206, etc.). An important criterion in ICP–MS is the ability to measure a weak signal intensity at mass, M, from an adjacent major peak, i.e. $M + 1$ or $M - 1$ (see Figure 6.1). This is termed *abundance sensitivity*. As you might imagine, this is an important factor in ICP–MS, where ultra-trace analysis is being carried out against a background of major element impurities (or the sample matrix). Values of up to 10^6 are achievable in quadrupole ICP–MS instruments.

6.4.2 Sector-Field Mass Spectrometer

The lack of resolution for a quadrupole instrument (i.e. limited to unit-mass resolution) makes it impossible to separate interferences (see Section 6.6) which coincide at the same nominal mass. A spectral resolution of $\approx 10\,000$ would

Table 6.2 Examples of the resolution required to separate ions of similar intensity. Adapted from Jakubowski, N. *et al.* [6]

Analyte ion	Interfering ion	Resolution[a]
$^{24}Mg^+$	$^{12}C_2^+$	1605
$^{28}Si^+$	$^{14}N_2^+$	958
$^{28}Si^+$	$^{12}C^{16}O^+$	1557
$^{44}Ca^+$	$^{12}C^{16}O_2^+$	1281
$^{51}V^+$	$^{35}Cl^{16}O^+$	2572
$^{52}Cr^+$	$^{40}Ar^{12}C^+$	2375
$^{54}Fe^+$	$^{40}Ar^{14}N^+$	2088
$^{56}Fe^+$	$^{40}Ar^{16}O^+$	2502
$^{63}Cu^+$	$^{40}Ar^{23}Na^+$	2790
$^{64}Zn^+$	$^{32}S^{16}O_2^+$	1952
$^{75}As^+$	$^{40}Ar^{35}Cl^+$	7775
$^{80}Se^+$	$^{40}Ar_2^+$	9688

[a] Resolution $= M/\Delta M$.

remove most of the interferences from polyatomic species, but much higher resolution is required for isobaric interferences (Table 6.2). In order to separate at high resolution, a mass spectrometer capable of much higher resolution is required. These mass spectrometer instruments are normally based on double-focusing, which require a magnetic and an electric-sector field. The use of an ICP with a high-resolution mass spectrometer was first reported in 1989 [7, 8]. A schematic diagram of one such instrument is shown in Figure 6.7. This high-resolution mass analyser consists of an electrostatic analyser (ESA) and a magnetic analyser. In addition to both analysers, the important aspect of the high-resolution instrument is the use of narrow entry and exit slits which control the number of ions passing through to the detector at any one time. This is known as a *high-resolution mass analyser* as it is able to focus both the energy and *m/z* ratio. Ions from the ICP pass through a narrow slit, hence allowing only ions correctly aligned on a particular axial plane to pass through, so resulting in a narrow beam of ions all travelling parallel to each other. The electrostatic analyser consists of two curved plates applied with a DC voltage, thus allowing the inner plate (negative polarity) to attract positively charged ions, while the outer plate (positive polarity) repels the ions. The ion beam then passes between the two plates and is both focused and curved through an angle of $40°$. Since only ions with a narrow range of kinetic energies (see Equation (6.2) below) are able to pass through the ESA, the latter thus forms an effective energy 'filter', so allowing ions of all masses to pass into the magnet analyser. In the energy field of the magnet, the ions are separated by their *m/z* ratios, such that ions of different masses follow different circular

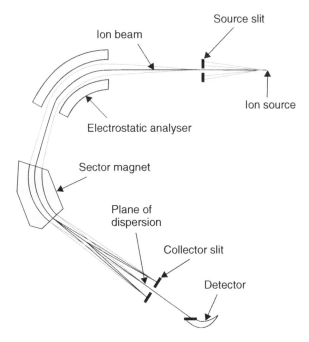

Figure 6.7 Schematic diagram of the layout of a high-resolution, double-focusing mass spectrometer. From 'Promotional Literature' published by Elemental Analysis, Inc., and reproduced with permission.

trajectories. By setting the field strength of the magnet, it is possible to select only those ions with a specific m/z ratio. The ion beam then passes through a narrow slit situated at the focal point of the magnet (collector slit). However, the high cost of these instruments precludes their wide availability. Figure 6.8 shows the resolution achievable by using this type of mass spectrometer for (a) the separation of iron from ArO^+ and (b) the determination of silicon in a steel sample.

6.4.3 Ion-Trap Mass Spectrometer

An ion-trap mass spectrometer consists of a cylindrical ring electrode and two end-cap electrodes (Figure 6.9). The top end-cap allows ions to be introduced into the trap where they are retained (i.e. ion-trapping). An RF voltage is applied to the ring electrode to stabilize and retain ions of different m/z ratios. Then, by increasing the applied voltage the paths of the ions of successive m/z ratios are rendered unstable. These ions then exit the trap via the bottom end-cap, prior to detection. Such a process can be described as 'being in a mass-selective instability mode'.

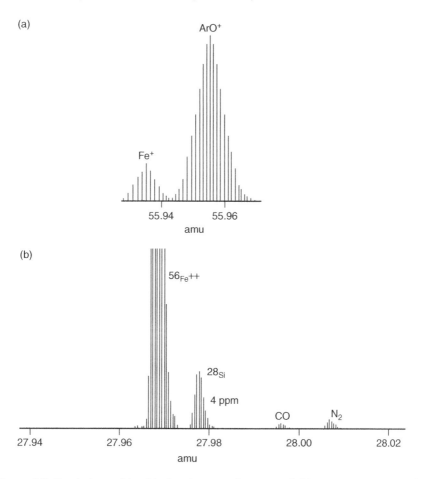

Figure 6.8 Resolution achievable by the use of a sector-field mass spectrometer for (a) the separation of iron from ArO^+, and (b) the determination of silicon in a sample of steel. (a) Reproduced by permission of Waters Corporation from 'Applications Literature' published by Micromass UK, Ltd. (b) From 'Applications Literature' published by VG Elemental. Reproduced by permission of Thermo Electron Corporation, Winsford, Cheshire, UK.

6.4.4 Time-of-Flight Mass Spectrometer

A time-of-flight (TOF) mass spectrometer does not rely on magnetic, electrostatic or an RF field to separate ions of different m/z ratios, but instead uses an applied accelerating voltage and the resultant different velocities of ions as the basis of its separation. As each ion of different m/z ratios has the same kinetic energy (see Equation (6.2)) but a different mass, it will travel through the time-of-flight spectrometer at a different velocity and as a result be separated.

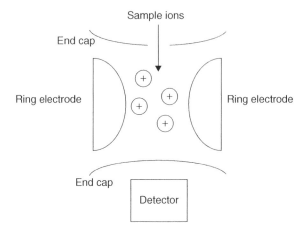

Figure 6.9 Schematic representation of the operation of an ion-trap mass spectrometer.

The kinetic energy is defined as follows:

$$\text{Kinetic energy (KE)} = mv^2/2 \tag{6.2}$$

where m is the mass of the ion and v its velocity.

As a result, the TOF mass spectrometer measures the time it takes for an ion to travel through the 'drift tube' from the source to the detector (Figure 6.10). The lighter ions, which are travelling faster, reach the detector before the heavier ions. To increase the drift path length, an ion reflector, or 'reflectron', is introduced into the mass analyser – this reverses the direction of flow of the ions, as well as doubling the flight path.

The disadvantage of the TOF mass spectrometer is that it requires a pulsed source of ions, whereas an ICP provides a continuous ion beam. One approach to circumnavigate this issue is by the addition of a quadrupole mass analyser (see Section 6.4.1) prior to the TOF mass spectrometer.

6.5 Detectors

The most common type of detector is the continuous dynode electron multiplier (Figure 6.11). The operating principles of this electron multiplier are similar to those of the photomultiplier tube (see Section 5.4.1), apart from the absence of dynodes. In addition, the electron multiplier must operate under vacuum conditions ($< 5 \times 10^{-5}$ torr). This device consists of an open tube with a wide entrance cone, with the inside of the tube being coated with a lead oxide semiconducting material. The cone is biased with a high negative potential (e.g. $-3\,\text{kV}$) at the entrance and held 'at ground' near the collector.

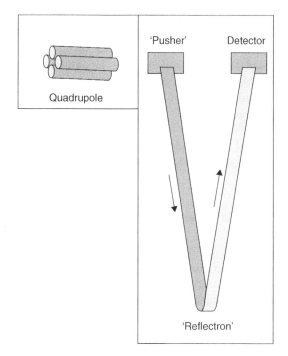

Figure 6.10 Schematic diagram of the layout of a time-of-flight mass spectrometer.

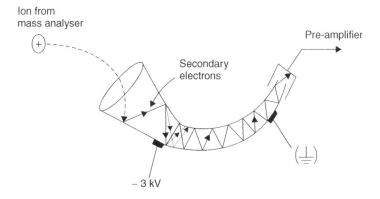

Figure 6.11 Schematic representation of the operating principles of the electron multiplier tube arrangement [5]. From Dean, J. R., *Atomic Absorption and Plasma Spectroscopy*, 2nd Edition, ACOL Series, Wiley, Chichester, UK, 1997. © University of Greenwich, and reproduced by permission of the University of Greenwich.

Any incoming positive ions, from the mass analyser, are attracted towards the negative potential of the cone. On impact, the positive ions cause one or more (secondary) electrons to be ejected. These secondary electrons are attracted towards the grounded collector within the tube. In addition, the initial secondary electrons can also collide with the surface coating, so causing further electrons to be ejected. This multiplication of electrons continues until all of the electrons (up to 10^8 electrons) are collected. Such a discrete pulse of electrons is further amplified exterior to the electron multiplier tube and recorded as a number of ion 'counts per second'. All electron multiplier tubes have a limited lifetime, determined by the total accumulated charge, which is monitored.

SAQ 6.4

Compare the operation of the electron multiplier tube with that of the photomultiplier tube.

The electron multiplier can also respond to photons of light from the ICP. For this reason, the detector can be mounted 'off-axis', the spectrometer can be 'off-axis' or a baffle can be located in the centre of the ion lens. Nevertheless, the electron multiplier tube is a sensitive detector for ICP–MS.

An alternative detector that can be used when ion currents exceed 10^6 ions s^{-1} is the *Faraday cup*. Ions impinging inside this detector are further amplified and counted. The advantage of this type of detector is its stability and freedom from 'mass bias', so allowing it to record highly accurate isotope ratio measurements. Its major disadvantage is its lack of sensitivity for trace elemental analysis. Another detector used in mass spectrometry is the '*Daly detector*'. This consists of a metal knob that emits secondary electrons when struck by an ion. The secondary (generated) electrons are accelerated onto a scintillator material which produces light – this can then be detected by the use of a photomultiplier tube (see section 5.4.1).

6.6 Interferences

While interferences in ICP–MS are not so prevalent as in ICP–AES, nevertheless some types of interferences do occur. The types of interferences can be broadly classified, according to their origin, into *spectral* (isobaric and molecular) and *non-spectral*. Spectral interferences can occur as a result of the overlap of atomic masses of different elements, i.e. isobaric interferences, and molecular processes. The latter can occur as a result of the acid(s) used to prepare the sample and/or the argon plasma gas (*polyatomics*). In addition, the formation of oxides, hydroxides and doubly charged species is possible. Non-spectral interferences or *matrix* interferences result in signal enhancement or depression with respect to the atomic mass.

6.6.1 Isobaric Interferences

These types of interferences are well characterized (Table 6.3) and as a result of the fact that $\sim 70\%$ of the elements in the periodic table have more than one isotope, can usually be avoided by selecting an alternative isotope. By considering Table 6.3, it is possible to identify situations in which potential problems are alleviated by the ability to select an alternative isotope.

For example, if you were going to analyse nickel in a stainless steel sample, it would be appropriate to select the most abundant isotope for nickel, which occurs at atomic mass 58 and is 67.9% abundant. However, iron (0.31% abundant) occurs at the same mass. Thus, in order to prevent isobaric interferences it may be necessary to select an alternative mass. At atomic mass 60, nickel is 26.2% abundant and so no interfering isotopes occur. One other point to consider is that by selecting a less-abundant isotope for nickel potentially leads to a lowering in sensitivity. This latter point may not be significant if nickel is present at a high enough concentration in the steel sample because of the inherent sensitivity of the technique.

SAQ 6.5

Can you identify a situation in which an alternative mass would be needed for the determination of zinc in a nickel alloy?

SAQ 6.6

Can you identify a situation for titanium in which the only available alternative mass is one with an inherently low-percentage abundance?

Unfortunately, other situations exist which do not have such readily amenable solutions. Probably the best example of this is the determination of calcium (atomic mass 40). Unfortunately (for Ca), the ICP source consists of argon ions which have a mass coincidence at atomic mass 40 (Ar, 99.6% abundant). In this situation, no alternative mass exists for Ca which will provide any degree of sensitivity. The best available mass is 44 amu at which Ca has an abundance of 2.08%. An alternative approach is to use collision/reaction cell technology (see Section 6.6.3.2).

SAQ 6.7

Can you identify a situation for selenium in which no alternative mass is available?

6.6.2 Molecular Interferences

Molecular interferences derive from several origins and can be sub-divided into two different types, i.e. *polyatomics* and *doubly charged polyatomic interferences.*

Table 6.3 Isobaric interferences from Period 4 of the Periodic Table [5]. From Dean, J. R., *Atomic Absorption and Plasma Spectroscopy*, 2nd Edition, ACOL Series, Wiley, Chichester, UK, 1997. © University of Greenwich, and reproduced by permission of the University of Greenwich

Atomic mass	Element of interest (% abundance)	Interfering element (% abundance)
39	K (93.10)	—
40	Ca (96.97)	Ar (99.6)[a]; K (0.01)
41	K (6.88)	—
42	Ca (0.64)	—
43	Ca (0.14)	—
44	Ca (2.06)	—
45	Sc (100)	—
46	—	Ca (0.003); Ti (7.93)
47	—	Ti (7.28)
48	Ti (73.94)	Ca (0.19)
49	—	Ti (5.51)
50	—	Ti (5.34); V (0.24); Cr (4.31)
51	V (99.76)	—
52	Cr (83.76)	—
53	—	Cr (9.55)
54	—	Cr (2.38); Fe (5.82)
55	Mn (100)	—
56	Fe (91.66)	—
57	—	Fe (2.19)
58	Ni (67.88)	Fe (0.33)
59	Co (100)	—
60	Ni (26.23)	—
61	—	Ni (1.19)
62	—	Ni (3.66)
63	Cu (69.09)	—
64	Zn (48.89)	Ni (1.08)
65	Cu (30.91)	—
66	Zn (27.81)	—
67	—	Zn (4.11)
68	—	Zn (18.57)
69	Ga (60.40)	—
70	—	Zn (0.62); Ge (20.52)
71	Ga (39.60)	—
72	Ge (27.43)	—
73	—	Ge (7.76)
74	Ge (36.54)	Se (0.87)
75	As (100)	—
76	—	Ge (7.76); Se (9.02)
77	—	Se (7.58)

Table 6.3 (*continued*)

Atomic mass	Element of interest (% abundance)	Interfering element (% abundance)
78	Se (23.52)	Kr (0.35)
79	Br (50.54)	—
80	Se (49.82)	Kr (2.27)
81	Br (49.46)	—
82	—	Se (9.19); Kr (11.56)
83	Kr (11.55)	—
84	Kr (56.90)	Sr (0.56)[b]
85	—	—
86	—	Kr (17.37); Sr (9.86)[b]

[a] Not in Period 4 of the Periodic Table but included because of its origin from the plasma source.
[b] Not in Period 4 of the Periodic Table but included for completeness.

6.6.2.1 Polyatomic interferences

Polyatomic interferences are derived as a result of selected interactions between the element of interest and its associated aqueous solution, the plasma gas (Ar) or the types of acid(s) used in the preparation of the sample (Table 6.4). The type of acid-derived interferences considered in Table 6.4 are due to nitric, sulfuric, hydrochloric and phosphoric acids. It now becomes evident that the spectra obtained from a quadrupole MS instrument may be more complicated than anticipated. Very few elements in Period 4 of the Periodic Table are unaffected by some type of interference (either isobaric or polyatomic). It is likely that the polyatomic ions are formed within the interface where the ions are undergoing transfer from the atmospheric source to the mass spectrometer vacuum. Nevertheless, they provide an unwelcome addition to the interpretation of mass spectra. As a consequence, the unsuspecting analyst may record a signal due to the presence of a polyatomic interference rather than the isotope (element) intended.

6.6.2.2 Doubly charged polyatomic interferences

In addition to the previously described polyatomic interferences derived essentially from the Ar plasma gas and the acids used, a further type of interference can be identified. This is due to the formation of doubly charged species. Remember that for a particular isotope (element), you are measuring its mass/charge ratio. If the charge, z, alters (normally $z = 1$), then the resultant mass/charge ratio will also change; e.g. for a charge of two ($z = 2$), the resultant mass/charge ratio will halve. Of particular concern for this type of interference is the formation of doubly charged species of Ce, La, Sr, Th and Ba.

Table 6.4 Potential polyatomic interferences derived from the element of interest and its associated aqueous solution, the plasma gas itself and the type of acid(s) used to digest or prepare the sample [5]. From Dean, J. R., *Atomic Absorption and Plasma Spectroscopy*, 2nd Edition, ACOL Series, Wiley, Chichester, UK, 1997. © University of Greenwich, and reproduced by permission of the University of Greenwich

Atomic mass	Element of interest (% abundance)	Polyatomic interference
39	K (93.10)	$^{38}Ar^1H^+$
40	Ca (96.97)	$^{40}Ar^+$
41	K (6.88)	$^{40}Ar^1H^+$
42	Ca (0.64)	$^{40}Ar^2H^+$
43	Ca (0.14)	—
44	Ca (2.06)	$^{12}C^{16}O^{16}O^+$
45	Sc (100)	$^{12}C^{16}O^{16}O^1H^+$
46	—	$^{14}N^{16}O^{16}O^+$; $^{32}S^{14}N^+$
47	—	$^{31}P^{16}O^+$; $^{33}S^{14}N^+$
48	Ti (73.94)	$^{31}P^{16}O^1H^+$; $^{32}S^{16}O^+$; $^{34}S^{14}N^+$
49	—	$^{32}S^{16}O^1H^+$; $^{33}S^{16}O^+$; $^{14}N^{35}Cl^+$
50	—	$^{34}S^{16}O^+$; $^{36}Ar^{14}N^+$
51	V (99.76)	$^{35}Cl^{16}O^+$; $^{34}S^{16}O^1H^+$; $^{14}N^{37}Cl^+$; $^{35}Cl^{16}O^+$
52	Cr (83.76)	$^{40}Ar^{12}C^+$; $^{36}Ar^{16}O^+$; $^{36}S^{16}O^+$; $^{35}Cl^{16}O^1H^+$
53	—	$^{37}Cl^{16}O^+$
54	—	$^{40}Ar^{14}N^+$; $^{37}Cl^{16}O^1H^+$
55	Mn (100)	$^{40}Ar^{14}N^1H^+$
56	Fe (91.66)	$^{40}Ar^{16}O^+$
57	—	$^{40}Ar^{16}O^1H^+$
58	Ni (67.88)	—
59	Co (100)	—
60	Ni (26.23)	—
61	—	—
62	—	—
63	Cu (69.09)	$^{31}P^{16}O_2^+$
64	Zn (48.89)	$^{31}P^{16}O_2^1H^+$; $^{32}S^{16}O^{16}O^+$; $^{32}S^{32}S^+$
65	Cu (30.91)	$^{33}S^{16}O^{16}O^+$; $^{32}S^{33}S^+$
66	Zn (27.81)	$^{34}S^{16}O^{16}O^+$; $^{32}S^{34}S^+$
67	—	$^{35}Cl^{16}O^{16}O^+$
68	—	$^{40}Ar^{14}N^{14}N^+$; $^{36}S^{16}O^{16}O^+$; $^{32}S^{36}S^+$
69	Ga (60.40)	$^{37}Cl^{16}O^{16}O^+$
70	—	$^{35}Cl_2^+$; $^{40}Ar^{14}N^{16}O^+$
71	Ga (39.60)	$^{40}Ar^{31}P^+$; $^{36}Ar^{35}Cl^+$
72	Ge (27.43)	$^{37}Cl^{35}Cl^+$; $^{36}Ar^{36}Ar^+$; $^{40}Ar^{32}S^+$
73	—	$^{40}Ar^{33}S^+$; $^{36}Ar^{37}Cl^+$
74	Ge (36.54)	$^{37}Cl^{37}Cl^+$; $^{36}Ar^{38}Ar^+$; $^{40}Ar^{34}S^+$
75	As (100)	$^{40}Ar^{35}Cl^+$
76	—	$^{40}Ar^{36}Ar^+$; $^{40}Ar^{36}S^+$

Table 6.4 (*continued*)

Atomic mass	Element of interest (% abundance)	Polyatomic interference
77	—	$^{40}Ar^{37}Cl^+$; $^{36}Ar^{40}Ar^1H^+$
78	Se (23.52)	$^{40}Ar^{38}Ar^+$
79	Br (50.54)	$^{40}Ar^{38}Ar^1H^+$
80	Se (49.82)	$^{40}Ar^{40}Ar^+$
81	Br (49.46)	$^{40}Ar^{40}Ar^1H^+$
82	—	$^{40}Ar^{40}Ar^1H^1H^+$
83	Kr (11.55)	—
84	Kr (56.90)	—
85	—	—
86	—	—

SAQ 6.8

For barium (atomic masses, 130, 132, 134, 135, 136, 137 and 138), identify the atomic masses at which Ba^{2+} species will occur?

6.6.3 Remedies for Molecular Interferences

As described above, a range of polyatomic interferences can and do occur in ICP–MS. However, advances have occurred in ICP–MS technology which allow these interferences to be reduced or minimized. Two particular remedies are now discussed, i.e. the use of a *cold plasma* or *collision/reaction cells*. Operation under cold-plasma conditions requires no instrumental modifications, whereas the use of a collision or reaction cell requires modification of the mass analyser. Commercial systems are available with collision/reaction cells integral to the mass analyser.

6.6.3.1 Cold-Plasma Conditions

The use of cold-plasma conditions has been developed to reduce interferences from argon. 'Cold plasma' refers to the operation of the ICP under low power and high-central (injector) gas-flow-rate conditions. A typical cold plasma would operate at a power of 0.6 kW (compared to 1 kW) and an injector gas-flow-rate of $1.1 \, l \, min^{-1}$ (compared to $0.7 \, l \, min^{-1}$). It has been reported [9] that the use of these operating conditions results in a reduction of the Ar^+, ArH^+, ArO^+ and Ar_2^+ species, but an enhancement of H_3O^+ and NO^+. However, cold-plasma conditions could result in enhanced matrix suppression effects. Such effects can be compensated for by the use of an internal standard.

6.6.3.2 Collision and Reaction Cells

While collision and reaction cells have been used extensively for fundamental studies in ion–molecule chemistry, it is only recently that they have been applied to ICP–MS [10, 11]. The collision/reaction cell is normally located behind the sample/skimmer cone arrangement and before the mass analyser. The use of collision and reaction cells in ICP–MS allows for the following:

- neutralization of the most intense chemical ionization species
- interferent or analyte ion mass/charge ratio shifts

These processes are affected by the use of a range of reaction types, including the following:

- charge exchange
- atom transfer
- adduct formation
- condensation reactions

General forms of each of these reaction types, with selected examples, are presented in the following (A, analyte; B, reagent; C, interferent):

Charge exchange

$$\text{General form:} \qquad C^+ + B \longrightarrow B^+ + C \qquad (6.3)$$

Charge exchange allows the removal of, for example, the argon plasma gas ion interference and the resultant formation of uncharged argon plasma gas, which is not then detected.

$$\text{Example:} \qquad Ar^+ + NH_3 \longrightarrow NH_3{}^+ + Ar \qquad (6.4)$$

Atom transfer: proton transfer

$$\text{General form:} \qquad CH^+ + B \longrightarrow BH^+ + C \qquad (6.5)$$

Proton transfer can remove the interference from, for example, ArH^+. This results in the formation of neutral (uncharged) argon plasma gas which is then not detected.

$$\text{Example:} \qquad ArH^+ + H_2 \longrightarrow H_3{}^+ + Ar \qquad (6.6)$$

Atom transfer: hydrogen-atom transfer

$$\text{General form:} \qquad C^+ + BH \longrightarrow CH^+ + B \qquad (6.7)$$

Hydrogen-atom transfer has the ability to alleviate an interference by increasing the mass/charge ratio by one.

$$\text{Example:} \qquad Ar^+ + H_2 \longrightarrow ArH^+ + H \qquad (6.8)$$

Atom transfer: hydride-atom transfer

$$\text{General form:} \quad A^+ + BH \longrightarrow B^+ + AH \quad (6.9)$$

Hydride-atom transfer can remove an interference by forming the hydride of an element with no charge.

Adduct formation

$$\text{General form:} \quad A^+ + B \longrightarrow AB^+ \quad (6.10)$$

Adduct formation, with, for example, ammonia, NH_3, allows the mass/charge ratio to increase by 17 amu (atomic weight of $N = 14$ and atomic weight of $H = 1$).

$$\text{Example:} \quad Ni^+ + NH_3 \longrightarrow Ni^{+\bullet}NH_3 \quad (6.11)$$

Condensation reaction

$$\text{General form:} \quad A^+ + BO \longrightarrow AO^+ + B \quad (6.12)$$

The use of a condensation reaction, in common with adduct formation, has the ability to increase the mass/charge ratio. For example, creation of the oxide of the element will increase the mass/charge ratio by 16 amu (atomic weight of $O = 16$).

$$\text{Example:} \quad Ce^+ + N_2O \longrightarrow CeO^+ + N_2 \quad (6.13)$$

In order to illustrate the practical use of this approach involving collision/reaction cells, two case studies are presented in the following.

Case Study 1
As indicated in Section 6.6.1, an important isobaric interference prevents the determination of calcium when using ICP–quadrupole MS (both calcium and argon have their major abundant ions at 40 amu, i.e. 96.9% and 99.6% abundancies, respectively). This interference can be alleviated by using a charge-exchange reaction, namely:

$$Ca^+ + Ar^+ + H_2 \longrightarrow Ca^+ + Ar + H_2^+ \quad (6.14)$$

Both calcium and argon   Argon atom formed
ions coincide at *m/z* 40 amu which is NOT
          separated by the spectrometer

The addition of hydrogen gas into the collision/reaction cell will remove any Ar ions present in the plasma, but will not effect the Ca ions in the plasma. This approach can be used to determine calcium in an argon ICP.

In general, however, the use of charge exchange to remove an interference will only work if the reagent gas to be used has an ionization potential between the interferent ion (e.g. Ar^+) and analyte ion (e.g. Ca^+). The other reaction types are not reliant on the ionization potential of the reagent gas, but are dependent upon thermodynamic and kinetic factors.

Case Study 2
The argument for the use of a high-resolution mass spectrometer (see Section 6.4.2) was demonstrated by its ability to separate iron from ArO^+ in Figure 6.8(a) (both result in signals at 56 amu). An alternative approach would be to use an atom-addition reaction which allows an analyte shift to occur:

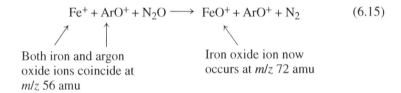

$$Fe^+ + ArO^+ + N_2O \longrightarrow FeO^+ + ArO^+ + N_2 \qquad (6.15)$$

Both iron and argon oxide ions coincide at m/z 56 amu

Iron oxide ion now occurs at m/z 72 amu

Note: while the removal of this interference at 56 amu is beneficial, it also creates another interference at 72 amu. Germanium has an isotope at 72 amu, which is 27.5% abundant. However, the major germanium isotope is ^{74}Ge, which is 36.35% abundant.

The use of collision/reaction cells can therefore lead to the 'chemical resolution' of isobaric and polyatomic interferences. However, it should also be noted that as they often result in different species being formed, this approach also has the potential to create other interferences. The major reagent gases used in ICP–MS [11] are as follows:

- collision gases, e.g. He
- charge-exchange gases, e.g. H_2, NH_3
- oxidation-reagent gases, e.g. O_2, N_2O
- reduction-reagent gases, e.g. H_2
- other reaction gases, e.g. CH_4

The application of this 'chemical-resolution' approach is very much under development at the present time and a whole range of applications are currently being developed. For the latest developments, the interested reader should consult the relevant scientific literature (see Chapter 8).

6.6.4 Non-Spectral Interferences: Matrix-Induced

Non-spectral interferences result from problems associated with the sample matrix. However, their origin within the system can vary. However, the resultant effect of non-spectral interferences is a loss of sensitivity for particular elements.

6.6.4.1 Problems Associated with the ICP

In the ICP, an incorrect choice of the appropriate sample-introduction device can lead to blockage problems in the nebulizer. Once this problem has been successfully negated, however, solids can still build-up on the sample cone of the ICP–MS interface. Both of these problems can lead to intermittent and erratic signal generation. Nevertheless, a number of remedies are possible and include the following:

- Choice of nebulizer, e.g. the use of a high-solids nebulizer (see Section 3.2).

- Aqueous dilution of the sample matrix (which may lead to decreased sensitivity of the analyte of interest) or the use of flow injection (see Section 3.5). Alternatively, the intermittent introduction of a sample with a high-salts content, followed by washing in dilute acid.

- Use of an internal standard. Application of the latter (addition of an element not in the sample at a fixed concentration to the standards and samples) will compensate for fluctuations in signal response.

- Matrix-matched standards. Matrix-matching of samples with calibration solutions may also be advantageous.

- The method of standard additions (see Section 1.4).

- 'On-line' coupling of a chromatographic separation technique to ICP–MS (see Section 3.5). The chromatographic separation technique may allow the preferential separation of the matrix from the element of interest by an ion-exchange process or, if the element has been chemically attached to an organic molecule, by reversed-phase high performance liquid chromatography or gas chromatography.

6.6.4.2 Problems Associated with the Mass Spectrometer

While mass-discrimination effects, which usually result in lower sensitivity, can be experimentally observed in the mass spectrometer, their exact mechanism is not known.

6.7 Isotope Dilution Analysis

The normal methods of calibration for atomic spectroscopy are external calibration and the method of additions (see Section 1.4). However, the use of a mass spectrometer provides an alternative approach, i.e. isotope dilution analysis (IDA). As any mass spectrometer is capable of measuring isotope ratios (ratios of the different isotopes of a single element), IDA uses this approach to provide a unique method of calibration. The essential feature of the IDA technique is that the element under investigation has more than one stable isotope; this applies to more than 70% of the elements in the Periodic Table. The basis of this approach is that isotope ratios before (the sample) and after 'spiking' (sample, plus spike) are measured. Then, by applying a mathematical solution, the concentration of the element in the sample can be determined. The main criterion for this approach is that the element spike must have an artificially enriched isotopic abundance, i.e. the isotopic composition of the element is different from that which would normally occur. This is illustrated in Figure 6.12, where Figure 6.12(a) shows the natural occurring isotopic composition of lead, i.e. lead that may be present in a sample, while Figure 6.12(c) shows the isotopic composition of an artificially enriched isotopic standard in which the abundance of ^{206}Pb has been increased by approximately 25%. The resultant mass spectrum will appear as shown in Figure 6.12(b). By measuring the isotopic ratio of the most abundant Pb isotope (^{208}Pb) with ^{206}Pb in the sample (208/206 Pb ratio ≈ 2.0) and comparing this with the resultant 208/206 Pb ratio in the sample plus spike (208/206 Pb ratio ≈ 1.0), an exact concentration of lead in the sample can be determined by using the following formula:

$$A = [x B_2(m_1/m_2) - B_1]/(z - zx/y) \qquad (6.16)$$

where A is the number of grams of element in the original sample, x the measured isotope ratio (^{208}Pb/^{206}Pb) in the spiked sample, B_1 the number of grams of ^{208}Pb in the enriched spike, i.e. the mass of total spike multiplied by the atom abundance of ^{208}Pb, B_2 the number of grams of ^{206}Pb in the enriched spike, i.e. the mass of total spike multiplied by the atom abundance of ^{206}Pb, m_1 the atomic weight of ^{208}Pb, m_2 the atomic weight of ^{206}Pb, z the fractional abundance of ^{208}Pb and y the isotope ratio (^{208}Pb/^{206}Pb) in the original sample.

It is worth noting that the values for B_1 and B_2 use the atom abundances for ^{208}Pb and ^{206}Pb, respectively, and not their % atom abundances as given in the following table.

Case Study 3
Calculate the fractional abundance, by weight, of ^{208}Pb.

The isotopic abundances of common lead are given in the following table (data from NIST Certificate of Analysis SRM 981).

Isotope	203.973	205.974	206.976	207.977
Atom abundance (%)	1.425	24.144	22.083	52.347

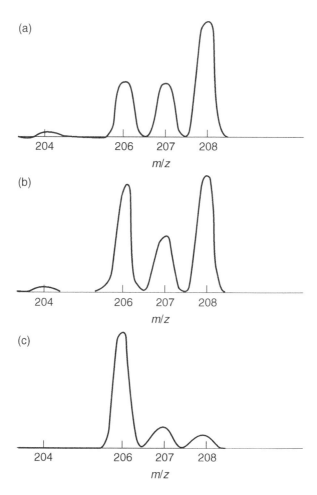

Figure 6.12 Mass spectra of isotopes of lead: (a) normal isotopic lead; (b) a mixture of normal isotopic lead, enriched with a [206]Pb 'spike'; (c) artificially enriched [206]Pb [5]. From Dean, J. R., *Atomic Absorption and Plasma Spectroscopy*, 2nd Edition, ACOL Series, Wiley, Chichester, UK, 1997. © University of Greenwich, and reproduced by permission of the University of Greenwich.

As the atomic weight of lead is 207.215, then the fractional abundance (by weight) of ^{208}Pb, z, can be calculated as follows:

$$z = (207.977 \times 52.347)/[(203.973 \times 1.425) + (205.974 \times 24.144)$$

$$+ (206.976 \times 22.083) + (207.977 \times 52.347)]$$

$$= 0.5254$$

It should be noted, however, that a quadrupole mass spectrometer may not measure the absolute isotopic composition of an element. It is likely, therefore, that the spectrometer will suffer from mass discrimination effects. If this is the case, it will be necessary to apply corrections to the measured isotopic ratios before carrying out the calculation when using isotope dilution analysis.

SAQ 6.9

The concentration of lead in an aqueous sample (100 ml) was 100 ng ml^{-1}. This sample was 'spiked' with an enriched ^{206}Pb standard (3.5 μg) by the addition of a 5 ml volume. After ICP–MS analysis, the following isotope ratios were found:

System	^{208}Pb/^{206}Pb
Original sample	2.1681
Sample plus enriched ^{206}Pb 'spike'	0.9521

Data for the isotopic composition of normal lead (from NIST Certificate of Analysis SRM 981) have been given earlier, while the corresponding data for the enriched ^{206}Pb 'spike' (from NIST Certificate of Analysis SRM 983) are given in the following table.

Isotope	203.973	205.974	206.976	207.977
Atom abundance (%)	0.0342	92.1497	6.5611	1.2550

By applying the principle of IDA, use the above data to confirm that the concentration of lead in the original sample is 100 ng ml^{-1}.

Isotope dilution analysis can be applied to all elements with more than one isotope. In principal, therefore, it can be used for multi-element analysis. However, in practice it is normally reserved for single-element determinations due to the high cost of the enriched stable isotopes and the increased time of each analysis. However, IDA does provide an ideal approach to internal standardization, in which one of the element's own isotopes acts as the *internal standard*. Use of the latter is known to improve the precision of the data. It is for this reason that IDA–ICP–MS is most commonly used by agencies involved in the production and certification of reference materials or in studies relating to nutrition, bioavailability and speciation.

6.8 Mass Spectral Interpretation

Mass spectral data obtained from an ICP–MS analysis are shown in Figure 6.13(a–f). By using the data presented in Table 6.5, it is possible to interpret the mass spectra. For the purposes of the following exercise, you can assume no molecular interferences. However, remember that isobaric interferences can and do occur.

SAQ 6.10

Using the data presented in Table 6.5, interpret the spectra shown in Figure 6.13(a–f).

Summary

The use of the inductively plasma for mass spectrometry is highlighted, with the fundamentals of mass spectrometry, as related to its combination with an inductively coupled plasma, being discussed. The major spectrometer designs for inductively coupled plasma–mass spectrometry are described. Particular emphasis

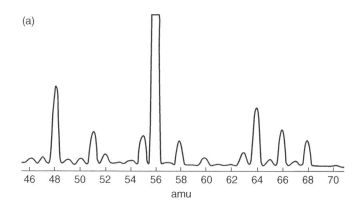

Figure 6.13 Set of mass spectra for peak identification (cf. SAQ 6.10) [5]. From Dean, J. R., *Atomic Absorption and Plasma Spectroscopy*, 2nd Edition, ACOL Series, Wiley, Chichester, UK, 1997. © University of Greenwich, and reproduced by permission of the University of Greenwich (*continued overleaf*).

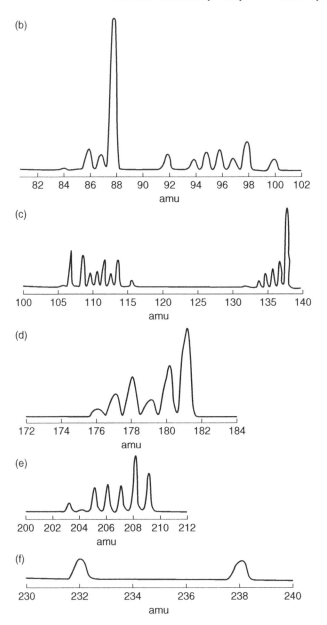

Figure 6.13 (*continued*).

Table 6.5 Relative abundances of selected naturally occurring isotopes [6], with data taken from Lide [12] (cf. SAQ 6.10).[a] From Dean, J. R., Atomic Absorption and Plasma Spectroscopy, 2nd Edition, ACOL Series, Wiley, Chichester, UK, 1997. © University of Greenwich, and reproduced by permission of the University of Greenwich

(a)

Element	amu												
	46	47	48	49	50	51	52	53	54	55	56	57	58
Ti	8.0	7.3	73.8	5.5	5.4								
V					0.2	99.8							
Cr					4.3		83.8	9.5	2.4				
Mn										100			
Fe											91.7	2.1	0.3
Ni									5.9		91.7	2.1	0.3
Ni													68.1

Element	amu											
	59	60	61	62	63	64	65	66	67	68	69	70
Co	100											
Ni		26.2	1.1	3.6		0.9						
Cu					69.2		30.8					
Zn						48.6		27.9	4.1	18.8		0.6

(b)

Element[a]	amu												
	82	83	84	85	86	87	88	89	90	91	92	93	94
(Kr)	11.6	11.5	57.0		17.3								
Rb				72.2		27.8							
Sr			0.5		9.9	7.0	82.6						
Y								100					
Zr									51.4	11.2	17.1		17.4
Nb												100	
Mo											14.8		9.3

(*continued overleaf*)

Table 6.5 (*continued*)

Element	amu					
	95	96	97	98	99	100
Zr		2.8				
Mo	15.9	16.7	9.5	24.1		9.6
Ru		5.5		1.9	12.7	12.6
Rh						
Pd						
Ag						
Cd						

(c)

Element	amu						
	101	102	103	104	105	106	107
Ru	17.1	31.6		18.6			
Rh			100				
Pd		1.0		11.1	22.3	27.3	
Ag							51.8
Cd						0.9	

Element	amu												
	108	109	110	111	112	113	114	115	116	117	118	119	120
Pd	26.5		11.7										
Ag		48.2											
Cd	0.9		12.5	12.8	24.1	12.2	28.7		7.5				
In						4.3		95.7					
Sn					1.0		0.6	0.4	14.5	7.7	24.2	8.6	32.6
Te													0.09

Element	amu												
	121	122	123	124	125	126	127	128	129	130	131	132	133
Sn		4.6		5.8									
Sb	57.4		42.6										
Te		2.6	0.9	4.8	7.1	18.9		31.7		33.9			
I							100						
Xe				0.1		0.1		1.9	26.4	4.1	21.2	27.0	
Cs													100
Ba										0.10		0.10	

(*continued*)

Table 6.5 (*continued*)

Element	amu								
	134	135	136	137	138	139	140	141	142
Xe	10.4		8.9						
Ba	2.4	6.6	7.9	11.2	71.7				
La					0.1	99.9			
Ce			0.2		0.3		88.4		11.1

(d)

Element[a]	amu										
	172	173	174	175	176	177	178	179	180	181	182
(Yb)	21.9	16.1	31.8		12.7						
Lu				97.4	2.6						
Hf			0.2		5.2	18.6	27.3	13.6	35.1		
Ta									0.0	99.9	
W									0.1		26.3

(e)

Element[a]	amu												
	200	201	202	203	204	205	206	207	208	209	210	211	212
(Hg)	23.1	13.2	29.9		6.9								
Tl				29.5		70.5							
Pb					1.4		24.1	22.1	52.4				
Bi										100			

(f)

Element	amu										
	230	231	232	233	234	235	236	237	238	239	240
Th			100								
U					0.0	0.7			99.3		

[a] Symbols in parentheses indicate elements that have incomplete percentage abundances (<100%).

is placed on the occurrence of interferences in mass spectrometry and the potential remedies for overcoming them. An important alternative strategy which is offered by the use of mass spectrometry, namely isotope dilution analysis, is also described.

References

1. Gray, A. L., *Proc. Soc. Anal. Chem.*, **11**, 182–183 (1974).
2. Houk, R. S., Fassel, V. A., Flesch, G. D., Svec, H. J., Gray, A. L. and Taylor, C. E., *Anal. Chem.*, **52**, 2283–2289 (1980).
3. Gray, A. L., *Spectrochim. Acta*, **40B**, 1525–1537 (1985).
4. Gray, A. L., *J. Anal. At. Spectrom.*, **1**, 403–405 (1986).
5. Dean, J. R., *Atomic Absorption and Plasma Spectroscopy*, 2nd Edition, ACOL Series, Wiley, Chichester, UK, 1997.
6. Jakubowski, N., Moens, L. and Vabhaecke, F., *Spectrochim. Acta*, **53B**, 1739–1763 (1998).
7. Morita, H., Ito, H., Uehiro, T. and Otsuka, K., *Anal. Sci.*, **5**, 609–610 (1989).
8. Bradshaw, N., Hall, E. F. H. and Sanderson, N. E., *J. Anal. At. Spectrom.*, **4**, 801–803 ((1989).
9. Douglas, D. J. and Tanner, S. D., 'Fundamental Considerations in ICP–MS', in *Inductively Coupled Plasma–Mass Spectrometry*, Montaser, A. (Ed.), Wiley-VCH, Weinheim, Germany, 1998, pp. 623–626.
10. Tanner, S. D., Baranov, V. I. and Bandura, D. R., *Spectrochim. Acta*, **57B**, 1361–1452 (2002).
11. Koppenaal, D. W., Eden, G. C. and Barinaga, C. J., *J. Anal. At. Spectrom.*, **19**, 561–570 (2004).
12. Lide, D. R. (Ed.), *CRC Handbook of Chemistry and Physics*, 73rd Edition, CRC Press, Boca Raton, FL, USA, 1992/1993, pp. 11-28–11-132.

Chapter 7

Selected Applications of Inductively Coupled Plasma Technology

Learning Objectives

- To appreciate the diversity of applications to which ICP technology can be applied.
- To consider the application of ICP–MS to document analysis.
- To appreciate the multi-element analysis capability of ICP–MS.
- To consider the application of ICP–AES and ICP–MS to coal analysis.
- To be able to compare the limits of detection of ICP–AES and ICP–MS with respect to coal analysis.
- To consider the application of ICP–MS in whole blood and urine analysis.
- To appreciate how the polyatomic interference on 75 amu between As and $^{40}Ar^{35}Cl^+$ and $^{40}Ca^{35}Cl^+$ can be corrected by the use of an empirical relationship when using an ICP with a quadrupole mass spectrometer.
- To appreciate how the polyatomic interference on 112 and 114 amu between Cd and Sn can be corrected by the use of an empirical relationship when using an ICP with a quadrupole mass spectrometer.
- To appreciate how the polyatomic interference on 75 amu can be eliminated by using a collision/reaction cell-ICP–MS system.
- To consider the application of ICP–MS in materials analysis.
- To appreciate the benefit of ICP–sector-field mass spectrometry (SFMS) in materials analysis.

Practical Inductively Coupled Plasma Spectroscopy   J. R. Dean
© 2005 John Wiley & Sons, Ltd

- To consider the application of ICP–MS in environmental analysis.
- To appreciate the role of isotope dilution analysis and its application to platinum group element analysis in roadside soil samples.
- To consider the application of ICP–MS in food analysis.
- To appreciate the importance of ICP–SFMS to overcome polyatomic interferences.
- To consider the application of ICP–MS in pharmaceutical analysis.
- To appreciate the importance of using liquid chromatography coupled to ICP–MS for problem-solving applications.

This chapter investigates the application of inductively coupled plasmas for elemental analysis of a range of sample types. These applications have been selected by the author on the basis of being either (a) of personal interest, or (b) representative of the type of analyses carried out. While hundreds of other similar examples could have been chosen to reflect the application of ICP in elemental analysis, the ones selected were considered appropriate by the author. (Note that this author has no links with any of the (co)-authors of the work selected.)

7.1 Forensic Science: Document Analysis

Document analysis is an important area of forensic evidence, particularly with respect to forgery and counterfeiting. One way of identifying the source of the paper is to characterize its elemental composition. This is particularly useful when examining successive pages of a multiple-page document to identify whether the paper is from a single source or different sources.

SAQ 7.1
Why might the elemental composition of paper vary?

Spence *et al.* [1] have used ICP–MS to characterize paper after microwave digestion with nitric acid and hydrogen peroxide. Printer/copy paper, i.e. white, A4 (210 × 297 mm) and 80 grams per square metre (gsm), was obtained from seventeen different sources (Table 7.1) and analysed by ICP–MS. The operating conditions for the ICP and data acquisition parameters for the MS are given in Table 7.2. Paper samples were digested by using a microwave oven according to the procedure shown in Figure 7.1. In order to ensure that the polytetrafluoroethylene (PTFE) microwave vessels were clean, they were previously acid washed and then rinsed with 'ultrapure' water. This was also carried out between each paper digest. Five replicates were digested per paper type. A procedural blank was also prepared per microwave digestion (i.e. six sample vessels were in the microwave digestion system during its operating cycle). Calibration was achieved

Table 7.1 Origin of paper for document analysis [1]

Country of origin	Number of different papers from the source
Australia	6
Austria	1
European Union	1
Finland	2
Indonesia	2
Japan	1
New Zealand	1
Scotland	1
Thailand	1
USA	1

Table 7.2 ICP–MS operating conditions and data acquisition parameters used for the analysis of paper [1]

ICP operating conditions	Details
RF power	1 kW
Outer gas-flow rate	15 l min^{-1}
Intermediate gas-flow rate	0.8 l min^{-1}
Nebulizer gas-flow rate	0.81–0.85 l min^{-1}

Mass spectrometer: data acquisition	Details
Mode	Peak hopping
Dwell time	50 ms
Points	1 per spectral peak
Sweeps	7 per reading
Number of replicates	1

by using an instrument response curve, over the mass range 3–238 amu, by the use of an external standard containing fifty eight elements, all at a concentration of 100 ng ml^{-1}. The external standard was re-analysed itself after each set of digests had been analysed, i.e. five digests of each paper plus blank.

Evaluation of the effectiveness of this approach for forensic document analysis required the identification of elements that could satisfy the following criteria:

- be well above their detection limits

- be homogeneously distributed throughout the paper

- not be subject to isobaric or molecular interferences

- have significant variations from one paper source to another

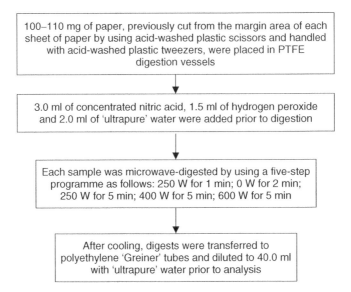

Figure 7.1 Procedure used for the microwave-oven digestion of paper [1].

It was found, based on the above criteria, that twenty three elements could be determined in the paper samples. However based on the criteria described, the above experiments were carried out to assess the suitability of using all twenty three elements. After evaluation, it was found that only nine elements were suitable as discriminators, namely, aluminium, cerium, lanthanum, magnesium, manganese, sodium, strontium, thorium and zirconium. Table 7.3 shows the typical elemental composition for a single paper (from Australia). Statistical evaluation of all of the data (student's t-test, 99% confidence) identified that all seventeen papers could be distinguished based on two elements only,

Table 7.3 Typical elemental composition of an Australian paper [1]

Element	Concentration ($\mu g\ g^{-1}$)
Na	403 ± 12
Mg	1006 ± 15
Al	1270 ± 50
Mn	74.8 ± 1.5
Sr	41.7 ± 1.3
Zr	4.09 ± 0.14
La	0.822 ± 0.017
Ce	1.46 ± 0.04
Th	0.491 ± 0.013

i.e. manganese and strontium content. Therefore, the use of ICP–MS offers an attractive proposition for the identification of different paper types, particularly for documents with successive pages.

7.2 Industrial Analysis: Coal

The use of coal for heating and electric production raises issues concerning its impact on the environment. As coal is composed of organic matter and minerals, it is rich in major, minor and trace elements. This example [2] focuses on the development of a microwave-digestion approach for the dissolution of coal prior to analysis by ICP–AES and ICP–MS. Three certified reference materials (CRMs) were analysed to assess the accuracy of the approach. The CRMs were obtained from the USA National Institute of Science and Technology, i.e. SRM 1632c, a bituminous coal, plus two sub-bituminous coals from the European Community Bureau of Reference (BCR 180) and the South African Bureau of Standards (SARM 19). The ICP–AES analysis was carried out on an axially viewed plasma with a concentric nebulizer, while ICP–MS was performed by using a quadrupole-based system. Six major elements (Al, Ca, K, Fe, Mg and Na) and fifteen minor/trace elements (As, Ba, Be, Cd, Co, Cr, Cu, Ga, Mn, Ni, Pb, Se, Sr, V and Zn) were analysed by using ICP–AES, while eighteen minor/trace elements (As, Ba, Be, Cd, Co, Cr, Cs, Cu, Ga, Li, Mn, Ni, Pb, Rb, Se, Sr, V and Zn) were analysed by ICP–MS.

The limits of quantitation (LOQs) (based on ten times the standard deviation of the blank signal multiplied by the dilution factor) were calculated for both ICP–AES and ICP–MS for the elements investigated (Table 7.4).

DQ 7.1

What can you infer from the LOQ data?

Answer

Essentially, it can be seen (Table 7.4) that the LOQs are predominantly lower for ICP–MS (up to 1–2 orders of magnitude) when compared to ICP–AES.

The coal CRMs were microwave-digested according to the procedure presented in Figure 7.2. This procedure did not require the use of hydrofluoric acid, which is both toxic and unsuitable for use with glass materials, e.g. volumetric flasks. The results obtained are shown in Table 7.5.

DQ 7.2

Comment on the results given in Table 7.5. Note that it is important to compare data between the different analytical techniques and with the certificates values.

Table 7.4 Limits of quantitation data (in $\mu g \ g^{-1}$) for elements in coal using ICP–AES and ICP–MS [2]

Element	ICP–AES		ICP–MS	
	Wavelength (nm)	LOQ[a]	Isotope (m/z)	LOQ[b]
Li	—	—	7	0.05
Be	313.11	0.04	9	0.05
V	290.88	0.1	51	0.04
Cr	267.72	0.5	53	0.5
Mn	257.61	0.06	55	0.09
Co	228.62	0.2	59	0.01
Ni	231.60	1.0	60	0.2
Cu	324.75	0.5	63	0.07
Zn	213.86	0.2	66	0.1
Ga	294.36	1.3	71	0.05
As	188.98	1.3	75	0.2
Se	196.03	2.6	82	1.0
Rb	—	—	85	0.004
Sr	407.77	0.01	88	0.006
Cd	228.80	0.2	111	0.05
Cs	—	—	133	0.006
Ba	455.40	0.03	137	0.08
Hg	194.17	0.7	202	0.7
Pb	220.35	1.5	208	0.06

[a]Dilution factor of 300.
[b]Dilution factor of 3000.

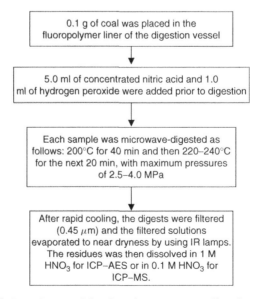

0.1 g of coal was placed in the fluoropolymer liner of the digestion vessel

↓

5.0 ml of concentrated nitric acid and 1.0 ml of hydrogen peroxide were added prior to digestion

↓

Each sample was microwave-digested as follows: 200°C for 40 min and then 220–240°C for the next 20 min, with maximum pressures of 2.5–4.0 MPa

↓

After rapid cooling, the digests were filtered (0.45 μm) and the filtered solutions evaporated to near dryness by using IR lamps. The residues was then dissolved in 1 M HNO_3 for ICP–AES or in 0.1 M HNO_3 for ICP–MS.

Figure 7.2 Procedure used for the microwave-oven digestion of coal [2].

Table 7.5 Analysis of Certified Reference Materials for coal from ICP–AES and ICP–MS after microwave-digestion using HNO$_3$/H$_2$O$_2$ (all data in μg g^{-1}) [2][a]

Element	SRM 1632c					BCR 180					SARM 19				
	Certified	ICP–AES	RSD (%)[b]	ICP–MS	RSD (%)[b]	Certified	ICP–AES	RSD (%)[b]	ICP–MS	RSD (%)[b]	Certified	ICP–AES	RSD (%)[b]	ICP–MS	RSD (%)[b]
Li	(8)	—	—	7.8	2.1	—	—	—	15.3	0.5	(37)	—	—	38.1	2.4
Be	(1)	0.92	4.3	0.94	4.2	—	0.69	4.5	0.73	2.2	2.8	2.45	2.8	2.61	1.9
Na	299 ± 6	287	1.0	—	—	(448)	453	1.6	—	—	2150 ± 70	1741	0.7	—	—
Mg	(384 ± 31)	304.4	0.5	—	—	(600)	631	0.9	—	—	1200 ± 110	1237	0.8	—	—
Al	(9150 ± 92)	8390	0.6	—	—	(12 400)	11948	0.6	—	—	39 000 ± 800	37 300	0.9	—	—
K	1100 ± 33	1146	0.9	—	—	(1200)	1127	1.8	—	—	1990	782	1.2	—	—
Ca	(1450 ± 304)	1120	0.6	—	—	—	3108	0.6	—	—	9930 ± 140	9363	0.1	—	—
V	(23.7 ± 0.5)	21.7	1.1	22.5	0.9	19.3 ± 0.6	18.4	0.6	17.7	1.0	35 ± 2	26.8	0.4	30.1	8.2
Cr	(13.7 ± 0.1)	12.3	1.0	13.5	2.1	(13.5)	12.6	0.3	12.8	4.0	50	43.5	0.3	48	0.5
Mn	(13.0 ± 0.5)	11.8	0.6	12.7	0.5	34.3 ± 1.1	33.8	0.4	37.2	0.4	157	138	0.3	—	—
Fe	(7350 ± 50)	7360	0.8	—	—	(11 700)	11470	0.4	—	—	12 250 ± 140	0.2	0.8	—	—
Co	3.5 ± 0.2	3.7	0.8	3.20	0.8	(3.3)	3.53	0.3	3.06	2.0	5.6	5.68	0.4	6.04	1.4
Ni	(9.3 ± 0.5)	10.0	0.7	10.5	2.1	—	8.6	0.5	8.9	0.2	16	14.4	0.2	15.9	2.9
Cu	(6.0 ± 0.2)	4.6	2.0	5.7	2.4	(9.1)	7.0	0.9	9.2	1.9	13 ± 1	10.1	0.8	12.4	1.0
Zn	12.1 ± 1.3	14.8	0.6	14.2	1.6	27.4 ± 1.1	34.1	0.3	28.9	1.6	12	22.4	0.4	13.1	4.0
Ga	(3)	3.8	12	3.7	1.4	—	4.0	4.8	3.7	2.2	14 ± 1	13.2	2.9	13.9	2.9
As	(6.2 ± 0.2)	5.4	6.1	5.8	1.9	4.23 ± 0.19	3.9	6.1	4.56	1.2	7 ± 1	7.0	4.3	6.8	4.8
Se	1.33 ± 0.03	—	—	1.43	28	1.32 ± 0.06	—	—	1.4	9.1	(1)	—	—	2.3	37
Rb	7.5 ± 0.3	—	—	6.58	0.3	—	—	—	6.97	0.9	9 ± 1	—	—	4.28	2.1
Sr	(63.8 ± 1.3)	50.3	0.5	—	—	(8.3)	78	0.2	—	—	126	107.8	0.8	—	—
Cd	(0.071 ± 0.007)	0.25	9.5	0.12	14	0.21 ± 0.01	0.29	4.4	0.26	5.8	—	—	—	0.23	9.9
Cs	(0.59 ± 0.01)	—	—	0.61	1.1	—	—	—	0.67	1.8	1.4	—	—	1.23	1.0
Ba	41.1 ± 1.6	34.3	0.8	—	—	(157)	123	1.0	—	—	304	253	0.7	—	—
Pb	(3.79 ± 0.08)	2.77	12	3.92	1.6	17.5 ± 0.5	13.2	2.3	17.4	0.1	20 ± 3	14.6	0.4	18.5	4.7

[a] Figures shown in brackets are reference values only.
[b] RSD, relative standard deviation.

Answer

Consistent results were obtained by both analytical techniques for the determination of As, Be, Cr, Co, Cu, Ga, Mn, Ni and V. Some significant anomalies were noted for Zn in SARM 19, Cd in SRM 1632c and Pb in SRM 1632c and BCR 180.

Inconsistent results were obtained by ICP–AES for the determination of Al, Ba, Ca, Cu, K, Fe, Na, Pb, V and Zn with respect to the certificate values, while ICP–MS showed fewer examples of poor accuracy, e.g. Mn and Rb.

7.3 Clinical/Biological Analysis: Whole Blood and Urine

The determination of As, Cd, Hg, Pb and Tl in whole blood and urine was assessed by using two different types of ICP–MS systems [3]. The first of these was a conventional system based on a quadrupole mass analyser (Table 7.6), while the second system had a reaction/collision cell positioned within the mass analyser but prior to the quadrupole mass spectrometer (Table 7.6). Bismuth, gallium and rhodium were used as internal standards. As arsenic suffers from a polyatomic interference at 75 amu, a correction is required when using a quadrupole-based system. The empirical equation to correct for the polyatomic interference when using the conventional ICP–MS system for arsenic determination is as follows:

$$\text{Arsenic} = I \ (m/z \ 75)0.042 \ 56 \times I \ (m/z \ 51) - 0.003 \ 32 \times I \ (m/z \ 43) \quad (7.1)$$

Table 7.6 ICP–MS operating conditions and data acquisition parameters used for the elemental analysis of whole blood and urine [3]

ICP operating conditions	Without reaction cell	With reaction cell
RF power	1175 W	1300 W
Outer gas-flow rate	15 l min^{-1}	15 l min^{-1}
Intermediate gas-flow rate	1.1 l min^{-1}	1.2 l min^{-1}
Nebulizer gas-flow rate	0.925 l min^{-1}	0.9 l min^{-1}
Reaction cell nebulizer gas-flow rate	—	0.75 l min^{-1}
Cell gas-flow rate	—	0.2 ml min^{-1}
RF cell voltage	—	0.85 V
Mass spectrometer: data acquisition	Without reaction cell	With reaction cell
Dwell time	100 ms	100 ms[a]
Sweeps per reading	15	15
Resolution	0.7 ± 0.1 amu	0.7 ± 0.1 amu

[a]200 ms for As.

where I is intensity. The above is based on the linear relationships between $^{16}O^{35}Cl^+$ (51 amu) and $^{40}Ar^{35}Cl^+$, and between $^{40}Ca^{35}Cl^+$ and $^{43}Ca^+$ (43 amu).

Other elemental equations, when using the conventional ICP–MS system, are as follows:

$$\text{Gallium} = I\ (m/z\ 69) \tag{7.2}$$

$$\text{Cadmium} = I\ (m/z\ 114) - 0.027\ 47 \times I\ (m/z\ 118) + I\ (m/z\ 111)$$
$$+ I\ (m/z\ 112) - 0.039\ 95 \times I\ (m/z\ 118) \tag{7.3}$$

(*Note*: this equation contains a correction for a tin interference)

$$\text{Rhodium} = I\ (m/z\ 103) \tag{7.4}$$

$$\text{Mercury} = I\ (m/z\ 199) + I\ (m/z\ 200)$$
$$+ I\ (m/z\ 201) + I\ (m/z\ 202) \tag{7.5}$$

(i.e. summation of major isotopes)

$$\text{Thallium} = I\ (m/z\ 205) + I\ (m/z\ 203) \tag{7.6}$$

(i.e. summation of major isotopes)

$$\text{Lead} = I\ (m/z\ 208) + I\ (m/z\ 206) + I\ (m/z\ 207) \tag{7.7}$$

(i.e. summation of major isotopes)

$$\text{Bismuth} = I\ (m/z\ 209) \tag{7.8}$$

For the collision/reaction cell-ICP–MS system, the relevant equations are as follows:

$$\text{Arsenic} = I\ (m/z\ 75) \tag{7.9}$$

$$\text{Gallium} = I\ (m/z\ 69) \tag{7.10}$$

$$\text{Cadmium} = I\ (m/z\ 110) + I\ (m/z\ 111) + I\ (m/z\ 113) \tag{7.11}$$

(i.e. summation of major isotopes)

$$\text{Rhodium} = I\ (m/z\ 103) \tag{7.12}$$

$$\text{Mercury} = I\ (m/z\ 199) + I\ (m/z\ 200)$$
$$+ I\ (m/z\ 201) + I\ (m/z\ 202) \tag{7.13}$$

(i.e. summation of major isotopes)

$$\text{Thallium} = I\ (m/z\ 205) + I\ (m/z\ 203) \tag{7.14}$$

(i.e. summation of major isotopes)

$$\text{Lead} = I \ (m/z \ 208) + I \ (m/z \ 206) + I \ (m/z \ 207) \tag{7.15}$$

(i.e. summation of major isotopes)

$$\text{Bismuth} = I \ (m/z \ 209) \tag{7.16}$$

The main interference to overcome was the polyatomic interference on 75 amu which is due to $^{40}Ar^{35}Cl^+$ and $^{40}Ca^{35}Cl^+$. This interference can be removed by using a collision/reaction cell system with the addition of H_2 which leads to the formation of ^{40}Ar, $^{1}H_3^+$ and $^{1}H^{35}Cl$. The hydrogen was introduced into the reaction/collision cell as a mixture, i.e. 5% H_2–95% Ar. The other four elements (Cd, Hg, Pb and Tl) were determined with the reaction cell vented.

Calibration standards were prepared in acidic aqueous solutions 'spiked' into a urine matrix. Samples of whole blood and urine were diluted with a mixture of 2.5% *t*-butanol, 0.5% HCl and 2 mg l^{-1} Au, plus internal standards. The detection limits were determined by using both ICP–MS instruments, employing three times the standard deviations of ten aqueous blank signals (Table 7.7). It can be seen from this table that the detection limits are up to fifteen times better when using the collision/reaction cell.

Conventional and collision/reaction cell-ICP–MS systems were then used for the analysis of a certified reference material (NIST, SRM 2670: toxic metals in freeze-dried urine; normal and elevated levels) and certified controls ('Bio-Rad' lypochek urine metals control and whole-blood samples, Levels 1–3). In addition, in-house whole blood and urine samples were also analysed. The results are shown in Table 7.8. It was concluded that As could be effectively determined when using the collision/reaction cell-ICP–MS system, i.e. the interferences from $^{40}Ar^{35}Cl^+$ and $^{40}Ca^{35}Cl^+$ are suppressed. Comparable results, with respect to certificate values, were obtained for all of the elements analysed.

Table 7.7 Detection limits (μg l^{-1}) for each element using conventional or collision/reaction cell-ICP–MS [3]

Element	Conventional ICP–MS[a]	Collision/reaction cell-ICP–MS[b]
As	0.72	1.04
Cd	1.51	0.13
Hg	2.26	0.17
Pb	0.20	0.38
Tl	0.08	0.01

[a] Dilution, 1:10.
[b] Dilution, 1:25.

Table 7.8 Analytical data obtained for certified reference materials and certified control samples (urine and whole blood) by conventional and collision/reaction cell (C/RC)-ICP–MS (in units of $\mu g\ l^{-1}$ ± standard deviation (SD)) [3]

Urine analysis

Element	'Bio-Rad' Level 2			NIST 2670 elevated			In-house urine	
	Certified	Conventional ICP–MS	C/RC-ICP–MS	Certified	Conventional ICP–MS	C/RC-ICP–MS	Certified	C/RC-ICP–MS
As	177 ± 45	168 ± 9	171 ± 11	480 ± 100	535 ± 30	548 ± 32	397 ± 30	385 ± 24
Cd	12.8 ± 2.5	12.9 ± 0.9	13.0 ± 0.7	88 ± 3	88 ± 6	87 ± 6	20.2 ± 2.0	20.3 ± 0.7
Hg	125 ± 37	154 ± 6	130 ± 43	105 ± 8	89 ± 15	66 ± 35	—	—
Pb	62.2 ± 12.5	74.3 ± 2.4	74.5 ± 1.5	109 ± 4	88 ± 15	101 ± 8	122 ± 10	128 ± 24
Tl	195 ± 49	212 ± 16	208 ± 7	—	—	—	493 ± 50	473 ± 19

Whole blood analysis

Element	In-house whole blood		'Bio-Rad' Level 1		'Bio-Rad' Level 2		'Bio-Rad' Level 3	
	Certified	C/RC-ICP–MS	Certified	C/RC-ICP–MS	Certified	C/RC-ICP–MS	Certified	C/RC-ICP–MS
As	172 ± 20	167 ± 6	—	—	—	—	—	—
Cd	18.5 ± 1.0	18.1 ± 1.5	—	—	—	—	—	—
Hg	74 ± 20	76 ± 16	—	—	—	—	—	—
Pb	172 ± 10	182 ± 22	78 ± 6	81 ± 12	284 ± 43	291 ± 4	505 ± 76	537 ± 11
Tl	86 ± 10	84 ± 4	—	—	—	—	—	—

7.4 Materials Analysis: Gadolinium Oxide

Gadolinium oxide is used in a whole range of high-technology applications, including the following:

- infrared-absorbing automotive glasses
- petroleum-cracking catalysts
- gadolinium–aluminium garnets
- microwave applications
- colour television phosphors
- optical glass manufacturing
- the electronics industry

Of particular importance in gadolinium oxide analysis is its rare-earth element composition. Samples of pure gadolinium oxide have been analysed by using ICP–sector-field mass spectrometry (SFMS) (Table 7.9) for sixteen elements, i.e. Sc, Y and the fourteen lanthanides [4]. Samples of pure gadolinium oxide (100 mg) were accurately weighed and dissolved in 50 vol% nitric acid (10 ml) at 50°C for 30 min. After dissolution, the samples were diluted to 100.0 ml with 1 vol% nitric acid solution, to therefore give concentrations of 1000 µg ml^{-1}. The sample solutions were then further diluted by a factor of 1000.

A calibration curve was constructed by using standards in the concentration range 0.1–10 ng ml^{-1} ($r^2 = 0.999$ (see Chapter 1) for the majority of elements). Detection limits were determined for all of the rare-earth elements, plus Sc and Y, and their values are given in Table 7.10.

Table 7.9 ICP–SFMS operating conditions and data acquisition parameters/type used for the elemental analysis of gadolinium oxide [4]

ICP operating conditions	Value
RF power	1.3 kW
Outer gas-flow rate	15 l min^{-1}
Intermediate gas-flow rate	0.9 l min^{-1}
Nebulizer gas-flow rate	1.2 l min^{-1}
Mass spectrometer: data acquisition	Type/value
Scan mode	Electric scan
Points per peak	20
Resolution	300

Table 7.10 Detection limits for rare-earth elements, scandium and yttrium, using ICP–SFMS [4]

Element	Detection limit (pg ml^{-1})[a]
Sc	10.3
Y	9.8
La	0.5
Ce	0.8
Pr	4.4
Sm	3.5
Nd	1.6
Tb	3.9
Dy	3.7
Ho	3.8
Er	2.9
Tm	3.3
Yb	4.6
Lu	3.1
Th	0.7
U	0.5

[a] Based on three times the standard deviations of ten consecutive measurements of a blank (1% HNO$_3$) solution.

SAQ 7.2

Based on the chemical symbols used in Table 7.10, identify the names of as many rare-earth elements as possible.

The analysis of two samples of gadolinium oxide was then carried out to determine the level of impurities. The results, shown in Table 7.11, are for impurities in a gadolinium oxide sample produced by the Instituto de Pesquisas Energeticas e Nucleares (IPEN) in Brazil and in a certified Johnson Matthey Company sample. It is noted (Table 7.11) that the results are in general agreement with the certificate values, with the exception of ^{175}Lu.

DQ 7.3

Can you suggest a potential spectral interference for ^{175}Lu?

Answer

The spectral interference for ^{175}Lu is from ^{158}Gd^{16}O^{1}H.

These same authors have also published similar findings for the determination of rare-earth elements in praseodymium oxide [5].

Table 7.11 Analysis of impurities in gadolinium oxide by ICP–SFMS [4][a]

Element	IPEN sample		Johnson Matthey Company sample	
	Certified	Measured	Certified	Measured
Sc	3.89 ± 0.03	4.23 ± 0.09	3.89 ± 0.03	3.67 ± 0.09
Y	2.37 ± 0.02	1.40 ± 0.03	2.37 ± 0.02	2.28 ± 0.06
La	4.98 ± 0.08	7.75 ± 0.26	4.98 ± 0.08	5.15 ± 0.17
Ce	7.99 ± 0.12	6.21 ± 0.23	7.99 ± 0.12	7.32 ± 0.25
Pr	5.85 ± 0.15	5.31 ± 0.15	5.85 ± 0.15	5.38 ± 0.20
Nd	11.9 ± 0.33	2.45 ± 0.05	11.9 ± 0.33	12.3 ± 0.46
Sm	11.6 ± 0.41	7.34 ± 0.19	11.6 ± 0.41	12.6 ± 0.57
Eu	6.45 ± 0.11	8.21 ± 0.38	6.45 ± 0.11	7.15 ± 0.26
Tb	14.9 ± 0.39	10.3 ± 0.31	14.9 ± 0.39	15.8 ± 0.57
Dy	15.7 ± 0.66	6.92 ± 0.26	15.7 ± 0.66	16.4 ± 0.75
Ho	8.85 ± 0.28	7.49 ± 0.37	8.85 ± 0.28	8.76 ± 0.31
Er	7.12 ± 0.39	6.44 ± 0.14	7.12 ± 0.39	6.92 ± 0.26
Tm	2.15 ± 0.08	3.04 ± 0.06	2.15 ± 0.08	1.94 ± 0.04
Yb	3.86 ± 0.07	4.53 ± 0.15	3.86 ± 0.07	4.73 ± 0.11
Lu	8.22 ± 0.31	15.7 ± 0.67	8.22 ± 0.31	7.35 ± 0.24

[a]All values in $\mu g\ g^{-1} \pm$ one standard deviation (SD) ($n = 5$).

7.5 Environmental Analysis: Soil

Emissions from the exhausts of road vehicles constitute one of the more common incidents of environmental pollution. The output from the internal combustion engine produces emissions of hydrocarbons, carbon monoxide and nitrogen oxides. The main approach adopted to reduce such emissions is via the use of catalytic converters. However, the subject of this case study is the emissions from the *catalytic converter* itself [6]. These converters are often based on platinum, palladium and rhodium, which act to catalyze the emissions into water vapour, carbon dioxide and nitrogen. Catalytic converters were first introduced in the 1970s. Initially, they were based on platinum only, for removing hydrocarbons and carbon monoxide. Subsequent developments now mean that a range of platinum group elements (PGEs) can be used, based on combinations of Pt, Pd, Ir and Rh. This case study [6], based in Austria, describes an initial attempt to determine the extent of the problem.

Roadside samples were gathered from a range of sites across Austria, with the sites being selected on the basis of the following:

- traffic density

- access to roadside

- elevation and gradient of the traffic line

Table 7.12 Soil-sampling sites and their characteristics [6]

Location[a]	Opened to traffic	Vehicles per day	Speed limit (km h^{-1})	Precipitation (mm year^{-1})
1[b]	—	—	—	838
2	1985	20 182	130	781
3	1972	22 072	130	1270
4	1970	56 679	130	514

[a]Location 1, Lungau; Location 2, S36 at Knittelfeld; Location 3, A14 at Rankweil; Location 4, A23 at Sudost-Tangente (all locations are in Austria).
[b]Reference site.

In addition, only grass-covered sites without any roadside barriers were selected for sampling. Relevant details of these sites are provided in Table 7.12. Location 1 (at Lungau) was selected as a reference site, being located in the Alpine region with minimal traffic. Samples of soil were collected at different distances from the roadside and at two depths (see Table 7.15 below for details). The collected soil samples were air-dried at ambient temperature and then sieved to < 2 mm prior to digestion. Three certified reference materials (BCR 723, WGB-1 and TDB-1) were selected to evaluate the accuracy of the approach. One CRM was a road dust (BCR 723, IRMM, Belgium) while the other two CRMs were geological reference materials (WGB-1, a gabbro and TDB-1, a basalt). Both were obtained from CCRMP-CANMT, Canada. Details of the high-pressure ashing procedure employed for the soil samples are given in Figure 7.3.

All of the samples/standards were analysed by using ICP–MS with isotope dilution calibration. Concentrations of the PGEs, except for Rh were determined by measurement of the following isotope ratios:

- ^{101}Ru/^{99}Ru
- ^{102}Ru/^{99}Ru
- ^{108}Pd/^{105}Pd
- ^{108}Pd/^{106}Pd
- ^{106}Pd/^{105}Pd
- ^{187}Re/^{185}Re
- ^{193}Ir/^{191}Ir
- ^{194}Pt/^{198}Pt
- ^{195}Pt/^{198}Pt

Isobaric interferences of ^{102}Pd on ^{102}**Ru**, ^{106}Cd on ^{106}**Pd**, ^{108}Cd on ^{108}**Pd** and ^{198}Hg on ^{198}**Pt** were corrected prior to ratio calculation. Rhodium was determined

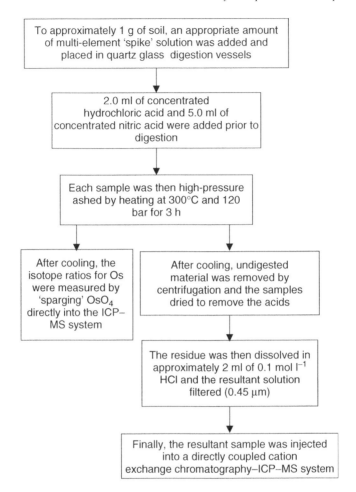

Figure 7.3 Procedure used for the high-pressure ashing of soil samples [6].

by its peak area in relation to the peak area of another PGE. The limits of detection and limits of determination are shown in Table 7.13.

The results obtained from analysis of the CRMs are shown in Table 7.14. All values reported are within the certificate value range (mean ± standard deviation (SD)). The procedure was then applied to analysis of the roadside samples, with the results obtained shown in Table 7.15.

DQ 7.4

What can you conclude about the results from location 1 in comparison to the natural background level reported in Table 7.15?

Table 7.13 Limits of detection and limits of determination (in units of ng g^{-1} ($n = 8$)) for PGEs and Re [6]

Element	Limit of detection	Limit of determination
Ru	0.014	0.058
Pd	0.015	0.046
Re	0.009	0.031
Os	0.004	0.012
Ir	0.010	0.050
Pt	0.064	0.176

Table 7.14 Analysis of certified reference materials (all values in ng g^{-1}) [6]

Element	BCR 723[a]		WGB-1[a]		TDB-1[a]	
	Certified value	Measured value ($n = 5$)	Certified value	Measured value ($n = 5$)	Certified value	Measured value ($n = 4$)
Ru	—	0.85 ± 0.29	(0.30)[b]	0.14 ± 0.02	(0.3)[b]	0.18 ± 0.003
Rh	12.8 ± 1.2	11.8 ± 1.4	(0.32)[b]	0.20 ± 0.03	(0.7)[b]	0.48 ± 0.06
Pd	6.0 ± 1.8	4.52 ± 0.23	13.90 ± 2.10	12.2 ± 1.70	22.4 ± 1.4	24.2 ± 2.47
Re	—	6.65 ± 0.09	—	1.18 ± 0.04	—	0.80 ± 0.03
Os	—	0.46 ± 0.01	—	0.59 ± 0.08	—	0.11 ± 0.01
Ir	—	0.53 ± 0.58	(0.33)[b]	0.20 ± 0.01	(0.15)[b]	0.07 ± 0.01
Pt	81.3 ± 3.3	82.4 ± 1.0	6.10 ± 1.60	4.75 ± 1.06	5.8 ± 1.1	4.98 ± 0.25

[a] Mean ± standard deviation (SD).
[b] 'Information values' only.

Answer

It can be seen that the levels monitored at location 1 (Lungau) are similar to the natural background concentration. This is not unexpected as this site was chosen to reflect an uncontaminated site (Alpine region with little traffic).

DQ 7.5

What can you conclude about the results from locations 2–4 in comparison to the natural background level reported in Table 7.15?

Answer

High levels of the catalytic converter elements, Rh, Pd and Pt, were found at locations 2–4. This corresponds to the locations chosen because of their high traffic volume. The highest values were reported at location 3

Table 7.15 Determination of PGEs and Re in roadside soil samples [6][a]

Sample	Distance from edge of road (m)	Depth (cm)	Ru	Rh	Pd	Re	Os	Ir	Pt
Natural background[b]	—	—	0.1	0.06	0.4	0.4	0.05	0.05	0.4
Location 1.1	—	0–5	0.02	0.03	0.36	0.13	0.04	0.04	0.28
Location 1.2	—	5–10	0.03	0.05	0.29	0.24	0.09	0.03	nd
Location 2.1	0.45	0–5	0.79	3.11	6.77	0.96	nd	0.44	32.4
Location 2.2	0.45	5–10	5.77	1.17	1.97	0.53	2.36	0.89	25.6
Location 2.3	2.55	0–5	0.15	1.96	1.54	0.23	0.14	0.09	12.7
Location 2.4	2.55	5–10	0.79	0.19	2.28	0.22	1.03	0.29	1.57
Location 2.5	10.50	0–5	0.12	0.17	0.90	0.05	0.08	0.12	1.13
Location 3.1	0.20	0–5	0.89	13.2	21.2	9.80	0.25	1.10	134
Location 3.2	0.20	5–10	0.29	12.6	24.5	5.41	0.07	0.15	102
Location 3.3	2.20	0–5	0.07	1.05	1.57	2.18	nd	0.11	9.14
Location 3.4	2.20	5–10	0.01	0.40	0.91	1.35	0.03	0.04	3.62
Location 3.5	10.20	0–5	0.02	0.48	0.79	0.28	0.03	nd	2.89
Location 4.1	0.65	0–5	0.55	3.39	6.00	0.88	0.04	0.09	38.9
Location 4.2	0.65	5–10	0.10	0.15	1.13	0.65	0.04	0.24	2.01
Location 4.3	2.65	0–5	0.23	1.52	6.41	0.71	0.04	0.18	27.2
Location 4.4	2.65	5–10	0.23	0.36	0.86	1.22	0.08	0.11	6.78
Location 4.5	6.50	0–5	0.07	0.74	1.75	0.78	0.06	0.10	7.04

[a] All values in ng g^{-1}; nd, not detected.
[b] Concentration in the continental crust [7].

at 0.2 m distances from the edge of the road, where the samples had maximum concentrations of 134 ng g^{-1} for Pt, 24.5 ng g^{-1} for Pd and 13.2 ng g^{-1} for Rh.

In general, these results have shown that the levels of the elements used in catalytic converters are on the increase. Further studies will therefore need to be carried out in order to ascertain the environmental impact of such elements on health.

7.6 Food Analysis: Milk Products

The analysis of trace and major elements in food products is important in terms of nutritional studies and toxicity assessment. Some elements are essential for nutritional purposes (Ca, Cu, Fe, Mg, Se, Zn, etc.), while others are toxic (Al, Cd, Cr, Hg, Mn, Ni and Pb). However, the use of an ICP with a quadrupole mass analyser does suffer from various interferences (see Section 6.6) when determining some of these elements. An alternative approach is to use an ICP with a double-focusing magnetic sector-field mass analyser (ICP–SFMS). This approach has been applied to the analysis of major, minor and trace essential elements, as well as toxic elements, in milk whey [8].

The ICP–SFMS system was operated under two different resolution ($m/\Delta m$) settings, i.e. 300 and 3000 (Table 7.16). This was necessary to allow separation of polyatomic interferences. It was determined that elements > 82 amu, i.e. Cd, Hg, Pb and Sr, were 'interference-free' in milk whey when using a resolution ($m/\Delta m$) of 300, whereas Al, Ca, Cr, Cu, Fe, Mg, Mn, Na, Ni, Se and Zn suffered from polyatomic interferences (Table 7.17).

Table 7.16 ICP–SFMS operating conditions and data acquisition parameter/type used for the elemental analysis of milk products [8]

ICP operating conditions	Value
RF power	1.3 kW
Outer gas-flow rate	14 l min^{-1}
Intermediate gas-flow rate	1.05 l min^{-1}
Nebulizer gas-flow rate	0.95 l min^{-1}
Mass spectrometer: data acquisition	Type/value
Mode	Electric scan
Number of scans	5 × 5
Sample time	0.01 s
Samples per peak	30 ($R = 3000$); 20 ($R = 300$)^a

^{a} R, resolution.

Table 7.17 Potential polyatomic interferences and resolutions required [8]

Isotope/element	Interference	Resolution required
^{27}Al	$^{13}C^{14}N$	1085
	$^{12}C^{14}N^{1}H$	919
	$^{12}C_{2}{}^{1}H^{2}H$	668
	$^{13}C^{12}C^{1}H_{2}$	720
^{42}Ca	$^{40}Ar^{2}H$	2350
^{44}Ca	$^{12}C^{16}O_{2}$	1281
^{45}Sc	$^{13}C^{16}O_{2}$	1207
	$^{12}C^{16}O_{2}{}^{1}H$	1080
^{52}Cr	$^{40}Ar^{12}C$	2376
^{53}Cr	$^{37}Cl^{16}O$	2626
^{55}Mn	$^{40}Ar^{15}N$	2248
	$^{40}Ar^{14}N^{1}H$	1559
	$^{39}K^{16}O$	2671
^{56}Fe	$^{40}Ar^{16}O$	2503
	$^{40}Ca^{16}O$	2479
^{57}Fe	$^{40}Ar^{16}O^{1}H$	1916
^{60}Ni	$^{44}Ca^{16}O$	3057
^{62}Ni	$^{23}Na_{2}{}^{16}O$	1343
	$^{38}Ar^{24}Mg$	3188
^{63}Cu	$^{40}Ar^{23}Na$	2792
	$^{31}P^{16}O_{2}$	1852
	$^{23}Na_{2}{}^{16}O^{1}H$	1194
	$^{15}N^{16}O_{3}$	1139
^{65}Cu	$^{31}P^{18}O^{16}O$	1622
	$^{32}S^{17}O^{16}O$	1695
	$^{33}S^{16}O_{2}$	1938
	$^{33}S^{32}S$	1939
^{82}Se	$^{1}H^{81}Br$	11 055

DQ 7.6

What resolution is required to analyse these elements?

Answer

Most elements, with the exception of ^{82}Se, can be analysed under a medium resolution $(m/\Delta m)$ of 3000. In addition, the internal standards used in this analysis were also investigated under medium-resolution conditions.

Samples of formula milk, UHT ('Longlife') cow's milk, human breast milk and infant formula milk were subjected to the following in order to precipitate and separate the fat and proteins from the whey: pH adjusted to 4.6, followed by centrifugation at 14 000 g for 30 min. The samples of whey were then diluted in 'ultrapure' water (1 + 4 wt% for minor and trace elements (Al, Cd, Cr, Cu, Fe, Hg, Mn, Ni, Pb, Se, Sr and Zn) and 1 + 1999 wt% for major elements (Ca, Mg and Na)). An internal standard stock solution containing Ga, In, Rh, Tl and Y at $10 \mu g\ l^{-1}$ concentration was added to each diluted milk whey sample.

The detection limits for each element (and selected isotopes) were determined based on three times the standard deviation of ten successive measurements of the blank and samples (Table 7.18). The accuracy was assessed by analysis of a certified reference material (BCR natural skim milk powder, CRM 063R). This CRM was re-constituted in 'ultrapure' water, as indicated above with respect to minor and trace elements, and major elements. The results obtained are shown in Table 7.19.

Table 7.18 Detection limits, using ICP–SFMS, for each element (isotope) in the original milk whey samples [8]

Isotope/element	Internal standard	Resolution used	Detection limit ($\mu g\ l^{-1}$)
^{23}Na	^{71}Ga	3000	25[a]
^{24}Mg	^{71}Ga	3000	1.6[a]
^{26}Mg	^{71}Ga	3000	0.6[a]
^{42}Ca	^{71}Ga	3000	13[a]
^{44}Ca	^{71}Ga	3000	1.8[a]
^{27}Al	^{71}Ga	3000	3.0
^{52}Cr	^{71}Ga	3000	0.03
^{53}Cr	^{71}Ga	3000	0.1
^{55}Mn	^{71}Ga	3000	0.05
^{56}Fe	^{103}Rh	3000	0.2
^{57}Fe	^{103}Rh	3000	1.5
^{60}Ni	^{71}Ga	3000	1.5
^{62}Ni	^{71}Ga	3000	2.0
^{63}Cu	^{103}Rh	3000	0.3
^{65}Cu	^{103}Rh	3000	0.3
^{66}Zn	^{103}Rh	3000	1.0
^{68}Zn	^{103}Rh	3000	1.5
^{82}Se	^{71}Ga	3000	1.0
^{86}Sr	^{89}Y	300	0.5
^{88}Sr	^{89}Y	300	0.2
^{110}Cd	^{115}In	300	0.03
^{111}Cd	^{115}In	300	0.04
^{200}Hg	^{205}Tl	300	0.5
^{202}Hg	^{205}Tl	300	0.4
^{206}Pb	^{205}Tl	300	0.2
^{208}Pb	^{205}Tl	300	0.15

[a]Detection limit in units of mg l^{-1}.

Table 7.19 ICP–SFMS analysis of natural skim milk powder (CRM, BCR 063R) [8]

Isotope/element	Certificate value	Concentration measured $(n = 5)$ (mg g^{-1})
^{23}Na	4.37 ± 0.04	4.3 ± 0.1
^{24}Mg	1.26 ± 0.04	1.27 ± 0.05
^{44}Ca	13.5 ± 0.2	13.3 ± 1.0
^{56}Fe	2.3 ± 0.4	2.1 ± 0.3
^{63}Cu	0.60 ± 0.03	0.5 ± 0.2
^{66}Zn	49 ± 1	49 ± 2
^{208}Pb	18.5 ± 2.1	19 ± 2
Isotope/element	Indicative value	Concentration measured $(n = 5)$ (µg g^{-1})
^{82}Se	0.13 ± 0.02	0.19 ± 0.03
Isotope/element	Information value	Concentration measured $(n = 5)$ (ng g^{-1})
^{27}Al	47 ± 9	47 ± 10
^{52}Cr	47 ± 8	52 ± 7
^{60}Ni	21 ± 6	nd^a
^{111}Cd	0.5–2.5	2.4 ± 0.4
^{202}Hg	0.19/0.36	nd^a

^{a}nd, not detected.

DQ 7.7

Comment on the accuracy of the data presented in Table 7.19.

Answer

Good agreement is found between the measured concentration and certificate values (including information values). Some discrepancies are noted, however, and include the 'indicative value' for selenium. This may be due to the interference on 82 amu from $^1H^{81}Br$. In addition, it was not possible to detect nickel and mercury in the sample. This procedure was then applied to the analysis of four types of milk product, namely, UHT cow's milk whey, cow's milk whey, human milk whey and formula milk whey (Table 7.20).

SAQ 7.3

What dilution was used for the major elements?

Table 7.20 Determination of major, minor and trace elements in milk whey samples [8]

Major elements	UHT cow's milk whey (mg l^{-1})	Cow's milk whey (mg l^{-1})	Human milk whey (mg l^{-1})	Formula milk whey (mg l^{-1})
^{23}Na	441 ± 9	429 ± 9	128 ± 4	467 ± 10
^{24}Mg	60 ± 2	84 ± 3	21 ± 2	59 ± 2
^{44}Ca	443 ± 8	700 ± 14	129 ± 6	571 ± 13

Minor/trace elements	UHT cow's milk whey (μg l^{-1})	Cow's milk whey (μg l^{-1})	Human milk whey (μg l^{-1})	Formula milk whey (μg l^{-1})
^{27}Al	nd	nd	nd	71 ± 5
^{52}Cr	nd	nd	nd	1.2 ± 0.1
^{55}Mn	5.6 ± 0.4	21 ± 1	6.1 ± 0.4	61 ± 1
^{56}Fe	44 ± 2	62 ± 3	250 ± 8	356 ± 17
^{60}Ni	nd	nd	nd	3.6 ± 0.2
^{63}Cu	7.3 ± 0.5	9.4 ± 0.9	118 ± 8	53 ± 4
^{66}Zn	0.87 ± 0.06^b	1.33 ± 0.09^b	1.23 ± 0.08^b	1.9 ± 0.1^b
^{82}Se	3.8 ± 0.2	4.1 ± 0.2	10 ± 5	8.7 ± 0.5
^{88}Sr	308 ± 23	284 ± 14	77 ± 5	283 ± 12
^{111}Cd	nd	nd	0.16 ± 0.05	0.28 ± 0.08
^{202}Hg	nd	nd	nd	nd
^{208}Pb	nd	0.9 ± 0.3	nd	1.2 ± 0.2

^{a} nd, not detected.
^{b} In units of mg l^{-1}.

SAQ 7.4

What was the concentration of the internal standard per element?

SAQ 7.5

Comment on the presence of essential elements (Ca, Cu, Fe, Mg, Se, Zn, etc.) in the milk whey samples.

SAQ 7.6

Comment on the presence of toxic elements (Al, Cd, Cr, Hg, Mn, Ni and Pb) in the milk whey samples.

7.7 Pharmaceutical Analysis

Pharmaceutical products need to be routinely screened for metals for two distinct reasons, namely:

- Metals are essential components of the pharmaceutical product, e.g. organometallic-based drugs for ulcer treatment, cancer treatment, etc.

- Metals can be present in the finished pharmaceutical product as a result of the following:

  - contamination during the manufacturing process

  - their use during the manufacturing process as catalysts

With respect to screening, the coupling of liquid chromatography with ICP–MS offers significant advantages and this approach has been recently reviewed [9]. Aspects covered include the following:

- Derivatization reactions with Cu and Fe to allow the determination of amines and carboxylic acids.

- Metabolism studies, involving the profiling of bromine-labelled peptides in blood plasma.

- Combinatorial chemistry – analysis of materials derived from single, solid-phase beads.

- Determination of protein phosphorylation on gel electrophoresis blots, by using laser-ablation ICP–MS.

Summary

The importance of inductively coupled plasma technology in providing valuable elemental information for a wide range of diverse applications is highlighted in this present chapter.

References

1. Spence, L. D., Baker, A. T. and Byrne, J. P., *J. Anal. At. Spectrom.*, **15**, 813–819 (2000).
2. Wang, J., Nakazato, T., Sakananishi, K., Yamada, O., Tao, H. and Saito, I, *Anal. Chim. Acta*, **514**, 115–124 (2004).
3. Nixon, D. E., Neubauer, K. R., Eckdahl, S. J., Butz, J. A. and Burrritt, M. F., *Spectrochim. Acta*, **B59**, 1377–1387 (2004).
4. Pedreira, W. R., da Silva Queiroz, C. A., Abrao, A. and Pimentel, M. M., *J. Alloys Compounds*, **374**, 129–132 (2004).

5. Pedreira, W. R., Sarkis, J. E. S., Rodrigues, I. A., Tomiyoshi, I. A. and da Silva Queiroz, C. A. and Abrao, A., *J. Alloys Compounds*, **323/324**, 49–52 (2001).
6. Fritsche, J. and Meisel, T., *Sci. Total Environ.*, **325**, 145–154 (2004).
7. Wedepohl, K. H., *Geochim. Cosmochim. Acta*, **59**, 1217–1232 (1995).
8. Martino, F. A. R., Sanchez, M. L. F. and Sanz Medel, A., *J. Anal. At. Spectrom.*, **15**, 163–168 (2000).
9. Marshall, P. S., Leavens, B., Heudi, O. and Ramirez-Molina, C., *J. Chromatogr. A*, **1056**, 3–12 (2004).

Chapter 8

Further Information: Recording of Data and Selected Resources

Learning Objectives

- To enable scientific information to be recorded in the laboratory.
- To suggest relevant additional resources that can be consulted, including books, journals and the Internet.

8.1 Recording of Information in the Laboratory

8.1.1 Introduction

All laboratory work must be carried out in accordance with safety regulations. These regulations, the Control of Substances Hazardous to Health (COSHH), in the UK, relate to an assessment of the hazards and risks associated with the use of chemicals and scientific equipment in the working (laboratory) environment.

When undertaking any laboratory work, it is essential to record information in a systematic manner relating to sample details (e.g. description of physical appearance of the sample), sample treatment required (e.g. drying, followed by grinding and sieving), the analytical technique (e.g. ICP–AES), calibration strategy (e.g. direct calibration using five standards, ranging in concentration from 0.5 to $5 \, \mu g \, ml^{-1}$) and the recording of results (e.g. signal readings for each standard and sample analysed). Included below are some examples of data sheets that could be used to ensure that all information is recorded in a systematic manner. These

Practical Inductively Coupled Plasma Spectroscopy   J. R. Dean
© 2005 John Wiley & Sons, Ltd

sheets are not intended to be totally comprehensive, and so may be amended as required.

The first data sheet (A) allows identification of any preliminary sample pre-treatment procedures that may be necessary prior to sample preparation. Data sheet (B) identifies the sample preparation requirements, while data sheets (C) and (D) contain, respectively, the instrumental requirements for ICP–AES and ICP–MS. (Data sheets (A), (C) and (D) are taken from Dean, J. R., *Methods for Environmental Trace Analysis*, AnTS Series. Copyright 2003. © John Wiley and Sons, Limited. Reproduced with permission.)

8.1.2 *Examples of Data Sheets*

Data Sheet A: Sample Pre-Treatment

- Sample description/identifier ....

- Source of sample ....

- Sample dried   yes/no
  Oven-dried (temperature and duration) ....
  Air-dried (temperature and duration) ....
  Other (specify) ....

- Grinding and sieving
  Grinder used (model/type) ....
  Particle size (sieve mesh size) ....

- Mixing of the sample
  Manual shaking   yes/no
  Mechanical shaking   yes/no
  Other (specify) ....

- Sample storage
  Fridge   yes/no
  Other (specify) ....

- Chemical pre-treatment
  pH adjustment   yes/no
  Addition of alkali (specify) .... or acid (specify) ....
  or buffer .... pH = ...

Data Sheet B: Sample Preparation for Inductively Coupled Plasma Analysis

- Sample weight(s): accurately weighed (record to four decimal places)

    Weight of sample and vessel ....g
    Weight of vessel ....g

    Weight of sample ....g

    Sample weights: 1 ....g
    Sample weights: 2 ....g
    Sample weights: 3 ....g

- Acid-digestion
    Vessel used..
    Hot-plate or other (specifiy) ....
    Temperature-controlled or not (specifiy) ....
    Type of acid used..
    Volume of acid used..ml

    Any other details..

- Other methods of sample decomposition, e.g. fusion, dry ashing, etc. (specify)..
    ....
    ....
    ....

- Sample dilution
    Specify, with appropriate units, the dilution factor involved ....
    ....
    ....
    ....

- Addition of an internal standard
    Specify..
    Added before digestion   yes/no
    Added after digestion   yes/no

- Sample and reagent blanks
    Specify..
    ....

Data Sheet C: Analytical Technique for Inductively Coupled Plasma–Atomic Emission Spectroscopy

- ICP characteristics
  Manufacturer ....
  Frequency .... Hz
  Power ....kW
  Observation height ....mm above load coil

- Argon gas-flow rates
  Outer gas-flow rate ....l min^{-1}
  Intermediate gas-flow rate ....l min^{-1}
  Injector gas-flow rate ....l min^{-1}

- Sample introduction method
  Nebulizer/spray chamber (specify) ....
  ....
  ....

- Spectrometer
  Simultaneous or sequential
  Element(s) and wavelength(s) ....
  ....

- Quantitation
  Peak height   yes/no
  Peak area   yes/no
  Method used   manual/electronic
  Internal standard ....
  External standard ....
  Calibration method   direct/standard additions
  Number of calibration standards ....
  Linear range of calibration ....

Data Sheet D: Analytical Technique for Inductively Coupled Plasma–Mass Spectrometry

- ICP characteristics
  Manufacturer ....
  Frequency .... Hz
  Power ....kW

- Argon gas-flow rates
  Outer gas-flow rate .... $l\,min^{-1}$
  Intermediate gas-flow rate .... $l\,min^{-1}$
  Injector gas-flow rate .... $l\,min^{-1}$

- Sample introduction method
  Nebulizer/spray chamber (specify) ....
  ....
  ....

- Mass spectrometer
  Element(s) and mass/charge ratio(s) ....
  ....
  Scanning or peak-hopping mode (specify) ....

- Quantitation
  Peak height   yes/no
  Peak area   yes/no
  Method used   manual/electronic
  Internal standard ....
  External standard ....
  Calibration method   direct/standard additions
  Number of calibration standards ....
  Linear range of calibration ....

8.2 Selected Resources

A range of relevant additional resources can be consulted to assist you with (a) a basic understanding of the techniques and approaches used for inductively coupled plasma spectroscopy, and/or (b) keeping 'up-to-date.'

8.2.1 Keeping 'Up-to-Date'

The most appropriate approach to keep 'up-to-date' is to consult relevant journals which publish research results on the use and applications of inductively coupled plasma spectroscopy. Some suggested journals are given in Table 8.1.

It is now common practice to find that publishers allow their journal contents to be accessed electronically (often for free – no charge). By 'electronically' signing-up for their e-mail 'alerting service', you can automatically be sent the latest contents pages for each journal selected. This allows the contents of each selected journal to be automatically forwarded to you directly at your e-mail address, thus enabling you to view the latest publications in a particular journal as they are published. This can be a valuable tool in obtaining the latest research information. A selection of major publishers of journals, plus their corresponding websites, are presented in Table 8.2.

Most journals are also available electronically on your desktop PC. This allows the full text to be read in either PDF or HTML formats. In PDF format, the article appears exactly like the print copy that you might find on a library shelf. In HTML format, the article will have hyperlinks to tables, figures or references (the latter may be further linked to their original source by using a 'reference linking' service). However, this facility requires the payment of a subscription fee, as is often the case for libraries in universities, industry or public organisations, etc.

8.2.2 Basic Understanding of Inductively Coupled Plasma Spectroscopy (and Related Issues)

The most appropriate resources for this are books that focus specifically on the topic of interest. As well as books that directly concern inductively coupled plasma spectroscopy, other important 'keywords' that might appear in the titles of books include the following:

- atomic spectroscopy

- inorganic mass spectrometry

- plasma spectroscopy

- instrumental chemical analysis

- analytical chemistry

Table 8.1 Alphabetical list of selected journals that publish articles (research papers, communications, critical reviews, etc.) on inductively coupled plasma spectroscopy

Journal	Publisher	Web address[a]
Analyst	The Royal Society of Chemistry	http://www.rsc.org/is/journals/current/analyst/anlpub.htm
Analytica Chimica Acta	Elsevier	http://authors.elsevier.com/JournalDetail.html?PubID=502681&Precis=DESC
Analytical Chemistry	American Chemical Society	http://pubs.acs.org/journals/ancham/
Journal of Analytical Atomic Spectrometry (JAAS)	The Royal Society of Chemistry	http://www.rsc.org/is/journals/current/jaas/jaaspub.htm
Journal of Environmental Monitoring (JEM)	The Royal Society of Chemistry	http://www.rsc.org/is/journals/current/jem/jempub.htm
Microchemical Journal	Elsevier	http://authors.elsevier.com/JournalDetail.html?PubID=620391&Precis=DESC
Spectrochimica Acta, Part B	Elsevier	http://authors.elsevier.com/JournalDetail.html?PubID=525437&Precis=DESC
Talanta	Elsevier	http://authors.elsevier.com/JournalDetail.html?PubID=525438&Precis=DESC
Trends in Analytical Chemistry	Elsevier	http://authors.elsevier.com/JournalDetail.html?PubID=502695&Precis=DESC

[a] As of October 2004. The products or material displayed are not endorsed by the author or the publisher of this present text.

Some specific books that may be useful for further in-depth study of the field of inductively coupled plasma spectroscopy are shown in Table 8.3. Other books that cover related issues, such as sampling, sample preparation and speciation studies, are shown in Tables 8.4–8.6, respectively. In addition, Table 8.7 suggests book titles that may be of interest in the wider context, for example:

- Reference materials and quality control in the laboratory
- Chemical (element) speciation and fractionation studies

Table 8.2 Selected major publishers of journals

Publisher	Web address[a]
The Royal Society of Chemistry	http://www.rsc.org
American Chemical Society	http://www.pubs.acs.org
Wiley	http://www.wiley.com
Elsevier	http://www.elsevier.com
CRC Press	http://www.crcpress.com
Springer-Verlag	http://springerlink.metapress.com

[a] As of October 2004. The products or material displayed are not endorsed by the author or the publisher of this present text.

Table 8.3 Specific books on Inductively Coupled Plasma Spectroscopy[a]

Lajunen, L. H. J. and Peramaki, P., *Spectrochemical Analysis by Atomic Absorption and Emission*, 2nd Edition, The Royal Society of Chemistry, Cambridge, UK, 2004.

Holland, J. G. and Tanner, S. D (Eds), *Plasma Source Mass Spectrometry: Applications and Emerging Technologies*, The Royal Society of Chemistry, Cambridge, UK, 2003.

Nolte, J., *ICP Emission Spectrometry: A Practical Guide*, Wiley-VCH, Weinheim, Germany, 2003.

Thomas, R., *Practical Guide to ICP–MS*, Marcel Dekker, New York, 2003.

Broekaert, J., *Analytical Atomic Spectrometry with Flames and Plasmas*, Wiley-Interscience, New York, 2002.

Beauchemin, D., Gregoire, D. C., Karanassios, V., Wood, T. J. and Mermet, J. M., *Discrete Sample Introduction Techniques for Inductively Coupled Plasma–Mass Spectrometry*, Elsevier, Amsterdam, The Netherlands, 2000.

Caruso, J. A., Sutton, K. L. and Ackley, K. L., *Elemental Speciation: New Approaches for Trace Element Analysis*, Elsevier, Amsterdam, The Netherlands, 2000.

Taylor, H., *Inductively Coupled Plasma–Mass Spectrometry: Practices and Techniques*, Academic Press, London, 2000.

Ebdon, L., Evans, E. H., Fisher, A. S. and Hill, S. J., *An Introduction to Analytical Atomic Spectrometry*, Wiley, Chichester, UK, 1998.

Montaser, A., *Inductively Coupled Plasma–Mass Spectrometry*, 2nd Edition, Wiley-VCH, Weinheim, Germany, 1998.

(continued overleaf)

Table 8.3 (*continued*)

Dean, J. R., *Atomic Absorption and Plasma Spectroscopy*, 2nd Edition, ACOL Series, Wiley, Chichester, UK, 1997.

Lobinski, R. and Marczenko, Z., *Spectrochemical Trace Analysis for Metals and Metalloids*, in *Wilson and Wilsons Comprehensive Analytical Chemistry*, Vol. XXX, Weber, S. G. (Ed.), Elsevier, Amsterdam, The Netherlands, 1996.

Evans, E. H., Giglio, J. J., Castillano, T. M. and Caruso, J. A., *Inductively Coupled and Microwave Induced Plasma Sources for Mass Spectrometry*, The Royal Society of Chemistry, Cambridge, UK, 1995.

Howard, A. G. and Statham, P. J., *Inorganic Trace Analysis. Philosophy and Practice*, Wiley, Chichester, UK, 1993.

Slickers, K., *Automatic Atomic Emission Spectroscopy*, 2nd Edition, Bruhlsche Universitatsdruckerei, Giessen, Germany, 1993.

Vandecasteele, C. and Block, C.B., *Modern Methods of Trace Element Determination*, Wiley, Chichester, UK, 1993.

Jarvis, K. E., Gray, A. L. and Houk, R. S., *Handbook of Inductively Coupled Plasma–Mass Spectrometry*, Blackie Academic and Professional, Glasgow, UK, 1992.

Lajunen, L. H. J., *Spectrochemical Analysis by Atomic Absorption and Emission*, The Royal Society of Chemistry, Cambridge, UK, 1992.

Holland, G. and Eaton, A. N., *Applications of Plasma Source Mass Spectrometry*, The Royal Society of Chemistry, Cambridge, UK, 1991.

Jarvis, K. E., Gray, A. L., Jarvis, I. and Williams, J. G., *Plasma Source Mass Spectrometry*, The Royal Society of Chemistry, Cambridge, UK, 1990.

Sneddon, J., *Sample Introduction in Atomic Spectroscopy*, Vol. 4, Elsevier, Amsterdam, The Netherlands, 1990.

Date, A. R. and Gray, A. L., *Applications of Inductively Coupled Plasma–Mass Spectrometry*, Blackie Academic and Professional, Glasgow, UK, 1989.

Moore, G. L., *Introduction to Inductively Coupled Plasma–Atomic Emission Spectroscopy*, Elsevier, Amsterdam, The Netherlands, 1989.

Thompson, M. and Walsh, J. N., *A Handbook of Inductively Coupled Plasma Spectrometry*, 2nd Edition, Blackie Academic and Professional, Glasgow, UK, 1989.

Adams, F., Gijbels, R. and van Grieken, R., *Inorganic Mass Spectrometry*, Wiley, New York, 1988.

Ingle, J. D. and Crouch, S. R., *Spectrochemical Analysis*, Prentice-Hall International, London, 1988.

Boumans P. W. J. M., *Inductively Coupled Plasma Emission Spectrometry*, Parts 1 and 2, Wiley, New York, 1987.

Montaser, A. and Golightly, D. W., *Inductively Coupled Plasmas in Analytical Atomic Spectrometry*, VCH Publishers, New York, 1987.

[a] Arranged in chronological order.

Table 8.4 Specific books on Sampling[a]

Conklin, A. R., *Field Sampling: Principles and Practices in Environmental Analysis*, Marcel Dekker, New York, 2004.

Bodger, K., *Fundamentals of Environmental Sampling*, Rowman & Littlefield Publishers, Lanham, MD, USA, 2003.

Muntau, H., *Comparative Evaluation of European Methods for Sampling and Sample Treatment of Soil*, Diane Publishing Company, Collingdale, PA, USA, 2003.

Popek, E. P., *Sampling and Analysis of Environmental Chemical Pollutants: A Complete Guide*, Academic Press, Oxford, UK, 2003.

Csuros, M., *Environmental Sampling and Analysis for Metals*, CRC Press, Boca Raton, FL, USA, 2002.

Csuros, M., *Environmental Sampling and Analysis: Laboratory Manual*, CRC Press, Boca Raton, FL, USA, 1997.

Ostler, N. K., *Sampling and Analysis*, Prentice Hall, New York, 1997.

Hess, K., *Environmental Sampling for Unknowns*, CRC Press, Boca Raton, FL, USA, 1996.

Keith, L. H., *Principles of Environmental Sampling*, Oxford University Press, Oxford, UK, 1996.

Keith, L. H., *Compilation of EPA's Sampling and Analysis Methods*, CRC Press, Boca Raton, FL, USA, 1996.

Harsham, K. D., *Water Sampling for Pollution Regulation*, Taylor and Francis, London, 1995.

Quevauviller, Ph., *Quality Assurance in Environmental Monitoring: Sampling and Sample Pretreatment*, Wiley, Chichester, UK, 1995.

Russell Boulding, J., *Description and Sampling of Contaminated Soils: A Field Guide*, CRC Press, Boca Raton, FL, USA, 1994.

Carter, M. R., *Soil Sampling and Methods of Analysis*, CRC Press, Boca Raton, FL, USA, 1993.

Baiuescu, G. E., Dumitrescu, P. and Gh. Zugravescu, P., *Sampling*, Ellis Horwood, London, 1991.

Keith, L. H., *Environmental Sampling and Analysis: A Practical Guide*, Lewis Publishers Inc., Chelsea, MI, USA, 1991.

---

[a] Arranged in chronological order.

- Statistics and chemometrics
- Laboratory safety

An alternative source of material is the Internet or World Wide Web. However, caution needs to be exercised in the use of the Internet in terms of the quality of the source material. It is recommended that only sources of known repute (e.g. professional bodies) are used. Some examples of suitable websites are given in Table 8.8.

Table 8.5 Specific books on Sample Preparation[a]

Dean, J. R., *Methods for Environmental Trace Analysis*, AnTS Series, Wiley, Chichester, UK, 2003.

Mitra, S. (Ed.), *Sample Preparation Techniques in Analytical Chemistry*, Wiley-Interscience, New York, 2003.

Dean, J. R., *Extraction Methods for Environmental Analysis*, Wiley, Chichester, UK, 1998.

Kingston, H. M. and Jassie, L. B., *Introduction to Microwave Sample Preparation*, ACS Professional Reference Book, American Chemical Society, Washington, DC, USA, 1988.

Bock, R., *A Handbook of Decomposition Methods in Analytical Chemistry*, International Textbook Company, London, 1979.

[a] Arranged in chronological order.

Table 8.6 Specific books on Speciation[a]

Quevauviller, Ph., *Method Performance Studies for Speciation Analysis*, The Royal Society of Chemistry, Cambridge, UK, 1997.

Van der Sloot, H. A., Heasman, L. and Quevauviller, Ph., *Harmonization of Leaching/Extraction Tests*, Elsevier, Amsterdam, The Netherlands, 1997.

Ure, A. M. and Davidson, C. M., *Chemical Speciation in the Environment*, Blackie Academic and Professional, Glasgow, UK, 1995.

Kramer, J. R. and Allen, H. E., *Metal Speciation. Theory, Analysis and Application*, Lewis Publishers Inc., Chelsea, MI, USA, 1988.

[a] Arranged in chronological order.

Table 8.7 Other useful books

Specific books on Validation, Quality Control and Reference Materials[a]

Chan, C. C., Lee, Y. C., Lam, H. and Zhang, X.-M., *Analytical Method Validation and Instrument Performance Verifications*, Wiley, Chichester, UK, 2004.

Ratliff, T. A., *The Laboratory Quality Assurance System. A Manual of Quality Procedures and Forms*, 3rd Edition, Wiley, Chichester, UK, 2003.

Barwick, V., Burke, S., Lawn, R., Roper, P. and Walker, R., *Applications of Reference Materials in Analytical Chemistry*, The Royal Society of Chemistry, Cambridge, UK, 2001.

Stoeppler, M., Wolf, W. R. and Jenks, P. J. (Eds.), *Reference Materials for Chemical Analysis*, Wiley, Chichester, UK, 2001.

Currell, G., *Analytical Instrumentation. Performance Characteristics and Quality*, AnTS Series, Wiley, Chichester, UK, 2000.

Prichard, F.E., *Quality in the Analytical Chemical Laboratory*, ACOL Series, Wiley, Chichester, UK, 1999.

Table 8.7 (*continued*)

Specific books on Chemical (Elemental) Speciation and Fractionation Studies[a]

Szpunar, J. and Lobinski, R., *Hyphenated Techniques in Speciation Analysis*, The Royal Society of Chemistry, Cambridge, UK, 2004.

Cornelius, R., Caruso, J., Crews, H. and Heumann, K., (Eds), *Handbook of Elemental Speciation: Techniques and Methodology*, Wiley, Chichester, UK, 2003.

Quevauviller, Ph., *Methodologies for Soil and Sediment Fractionation Studies*, The Royal Society of Chemistry, Cambridge, UK, 2002.

Ebdon, L., Pitts, L., Cornelius, R., Crews, H., Donard, A. F. X. and Quevauviller, Ph., *Trace Element Speciation for Environment, Food and Health*, The Royal Society of Chemistry, Cambridge, UK, 2001.

Quevauviller, Ph., *Method Performance Studies for Speciation Analysis*, The Royal Society of Chemistry, Cambridge, UK, 1998.

Van der Sloot, H. A., Heasman, L. and Quevauviller, Ph., *Harmonization of Leaching/Extraction Tests*, Elsevier, Amsterdam, The Netherlands, 1997.

Caroli, S., *Element Speciation in Bioinorganic Chemistry*, Wiley, Chichester, UK, 1996.

Ure, A. M. and Davidson, C. M., *Chemical Speciation in the Environment*, Blackie Academic and Professional, Glasgow, UK, 1995.

Kramer, J. R. and Allen, H. E., *Metal Speciation. Theory, Analysis and Application*, Lewis Publishers, Inc., Chelsea, MI, USA, 1988.

Specific books on Statistics and Chemometrics[a]

Chau, F. T., Leung, A. K. M, Liang, Y. Z. and Gao, J. B., *Chemometrics. From Basics to Wavelet Transform*, Wiley, Chichester, UK, 2004.

Brereton, R. G., *Chemometrics. Data Analysis for the Laboratory and Chemical Plant*, Wiley, Chichester, UK, 2003.

De Levie, R., *How to Use Excel in Analytical Chemistry and in General Scientific Data Analysis*, Cambridge University Press, Cambridge, UK, 2001.

Meier, P. C. and Zund, R. E., *Statistical Methods in Analytical Chemistry*, 2nd Edition, Wiley, Chichester, UK, 2000.

Miller, J. N. and Miller, J. C., *Statistics and Chemometrics for Analytical Chemistry*, 4th Edition, Prentice Hall, Harlow, UK, 2000.

Farrant, T. J., *Practical Statistics for the Analytical Chemist*, The Royal Society of Chemistry, Cambridge, UK, 1997.

Massart, D. L., Vandeginste, B. G. M., Buydens, L. M. C., de Jong, S., Lewi, P. J. and Smeyers-Verbeke, J., *Handbook of Chemometrics and Qualimetrics: Part A*, Elsevier, Amsterdam, The Netherlands, 1997.

Adams, M. J., *Chemometrics in Analytical Chemistry*, The Royal Society of Chemistry, Cambridge, UK, 1995.

(*continued overleaf*)

Table 8.7 (*continued*)

Specific books on Laboratory Safety[a]

Lewis, R. J., *Hazardous Chemicals. Desk Reference*, 5th Edition, Wiley, Chichester, UK, 2002.

Anon, *Safe Practices in Chemical Laboratories*, The Royal Society of Chemistry, London, 1989.

Anon, *COSHH in Laboratories*, The Royal Society of Chemistry, London, 1989.

Lenga, R. E. (Ed.), *Sigma-Aldrich Library of Chemical Safety Data*, 2nd Edition, Sigma-Aldrich Ltd, Gillingam, UK, 1988.

---

[a] Arranged in chronological order.

Table 8.8 Selected useful web sites

Organization	Web address[a]
American Chemical Society (ACS)	http://www.acs.org
International Union of Pure and Applied Chemistry (IUPAC)	http://iupac.chemsoc.org
Laboratory of the Government Chemist (LGC)	http://www.lgc.co.uk
National Institute of Standards and Technology (NIST) Laboratory	http://www.cstl.nist.gov
National Institute of Standards and Technology (NIST) 'WebBook'	http://webbook.nist.gov
The Royal Society of Chemistry (RSC)	http://www.rsc.org
Society of Chemical Industry (SCI)	http://sci.mond.org
United States Environmental Protection Agency (USEPA)	http://www.epa.gov
United States National Library of Medicine (USNLM)	http://chem.sis.nlm.nih.gov

[a] As of October 2004. The products or material displayed are not endorsed by the author or the publisher of this present text.

Summary

The recording of practical experimental details prior to, during and after trace elemental analysis is an essential component in laboratory work. Templates (data sheets) have been provided, as guidance, in order to ensure that the most appropriate information is recorded. In addition, details are given on the range of resources available, in both print and electronic format, to assist in the understanding of inductively coupled plasma technology and its application to trace elemental analysis.

Responses to Self-Assessment Questions

Chapter 1

Response 1.1

The following table shows the values that you should have obtained.

Quantity	m	μm	nm
3×10^{-7} m	0.000 000 3 m	$0.3\,\mu$m	300 nm
Quantity	mol l^{-1}	mmol l^{-1}	μmol l^{-1}
6.9×10^{-3} mol l^{-1}	0.006 9 mol l^{-1}	6.9 mmol l^{-1}	6900 μmol l^{-1}
Quantity	μg ml^{-1}	mg l^{-1}	ng μl^{-1}
2.80 ppm	$2.80\,\mu$g ml^{-1}	2.80 mg l^{-1}	2.80 ng μl^{-1}

Chapter 2

Response 2.1

Hydrofluoric acid is the reagent for dissolving silica-based materials. The silicates are converted to a more volatile species in solution, according to the following equation:

$$SiO_2 + 6HF = H_2(SiF_6) + 2H_2O$$

Practical Inductively Coupled Plasma Spectroscopy   J. R. Dean
© 2005 John Wiley & Sons, Ltd

Chapter 3

Response 3.1

Nd:YAG stands for neodymium yttrium aluminium garnet.

Chapter 5

Response 5.1

In order to calculate the wavelength, Equation (5.1) needs to be rearranged as follows:

$$\lambda = c/f$$

Therefore, as c (the velocity of light) has a value of 3.00×10^8 m s^{-1}, we can obtain the wavelength from the following (recall that the unit for frequency is Hz (or s^{-1})):

$$\frac{3 \times 10^8 \, \text{m s}^{-1}}{5 \times 10^{14} \, \text{s}^{-1}} = 6 \times 10^{-7} \, \text{m}$$

While the value from the calculation is 6×10^{-7} m, it is normal to represent this as 600 nm (or 0.000 000 600 m). This wavelength, 600 nm, occurs in the visible region of the electromagnetic spectrum.

Response 5.2

From Figure 5.4, the energy difference between these levels is 2.107 eV. By converting this value into joules (J), we obtain the following:

$$2.107 \times 1.602 \times 10^{-19} \, \text{J} = 3.375 \times 10^{-19} \, \text{J}$$

Recalling that $E = hc/\lambda$, the wavelength is therefore given by the following:

$$\lambda = \frac{6.626 \times 10^{-34} \, \text{J s} \times 3 \times 10^8 \, \text{m s}^{-1}}{3.375 \times 10^{-19} \, \text{J}} = 5.889(8) \times 10^{-7} \, \text{m, or } 589 \, \text{nm}$$

Response 5.3

Recalling that $E = hc/\lambda$, it is possible by using this equation to determine the energy of this spectral transition as follows:

$$E = \frac{6.626 \times 10^{-34} \, \text{J s} \times 3.00 \times 10^8 \, \text{m s}^{-1}}{589 \times 10^{-9} \, \text{m}} = 3.37 \times 10^{-19} \, \text{J}$$

or for 1 mole of photons, 3.37×10^{-19} J $\times 6.022 \times 10^{23}$ mol^{-1}, which gives:

$$E = 203\,544 \, \text{J mol}^{-1}$$

By substitution into the revised Boltzmann equation, we obtain:

$$N_1/N_0 = 2 \exp\ (-203\ 544\ \mathrm{J\,mol^{-1}}/8.314\ \mathrm{J\,K^{-1}mol^{-1}} \times 7000\ \mathrm{K})$$
$$N_1/N_0 = 2 \exp\ (-3.50)$$
$$= 0.06$$

Therefore, at a typical plasma temperature of 7000 K, 6% of the atoms are in the excited state.

Response 5.4

From Equation (5.6):

$$\Delta\lambda_N = \lambda^2 \Delta v/c$$
$$= (589 \times 10^{-9}\ \mathrm{m})^2(1/2.5 \times 10^{-9}\ \mathrm{s})/(3 \times 10^8\ \mathrm{m\,s^{-1}})$$
$$= [(3.469 \times 10^{-13})\mathrm{m}^2 \times (4 \times 10^8\ \mathrm{s^{-1}})]/3 \times 10^8\ \mathrm{m\,s^{-1}}$$
$$= 4.625 \times 10^{-13}\ \mathrm{m}$$

Therefore, $\Delta\lambda_N = 0.000\ 46\ \mathrm{nm}$

Response 5.5

From Equation (5.7):

$$\Delta\lambda_D = (2 \times 589 \times 10^{-9}\ \mathrm{m}/3 \times 10^8\ \mathrm{m\,s^{-1}})$$
$$\times \sqrt{(2 \times 8.314\ \mathrm{J\,K^{-1}mol^{-1}} \times 2500\ \mathrm{K}/23 \times 10^{-3}\ \mathrm{kg\,mol^{-1}})}$$

which becomes:

$$\Delta\lambda_D = (3.927 \times 10^{-15}\ \mathrm{s})\sqrt{(1\ 807\ 391\ \mathrm{J\,kg^{-1}})}$$

Recalling that $1\ \mathrm{J} = 1\ \mathrm{kg\,m^2\ s^{-2}}$, we obtain:

$$\Delta\lambda_D = (3.927 \times 10^{-15}\ \mathrm{s})\sqrt{(1807391\ \mathrm{kg\,m^2\ s^{-2}\ kg^{-1}})}$$

Then, by taking the square root of the term in the second bracket:

$$\Delta\lambda_D = (3.927 \times 10^{-15}\ \mathrm{s})(1344\ \mathrm{m^{2/2}\ s^{-2/2}})$$

which simplifies to:

$$\Delta\lambda_D = (3.927 \times 10^{-15}\ \mathrm{s})(1344\ \mathrm{m\,s^{-1}})$$

and therefore:

$$\Delta\lambda_D = 5.28 \times 10^{-12}\mathrm{m},\ \mathrm{or}\ 5.3\ \mathrm{pm},\ \mathrm{or}\ 0.0053\ \mathrm{nm}$$

Response 5.6

Below 190 nm, purged optics (using nitrogen) and a vacuum spectrometer are required due to the absorption of oxygen. However, the typical operation of most spectrometers is between 190 and 450 nm.

Response 5.7

A groove density of 1200 lines mm^{-1} is equivalent to one groove every 0.833×10^{-6} m (or $0.833 \mu m$). Therefore, $d = 0.833 \mu m$. The wavelength would then be 284 nm in the *first order*. In addition, at which wavelength would this same emission line occur in the *second order*? In this case, the value would be 142 nm.

Response 5.8

(a) From Equation (5.11), the resolution is given by:

$$R = 1 \times (1200\, mm^{-1} \times 52\, mm)$$
$$= 62\,400\ (notice\ no\ units)$$

(b) At a wavelength of 300 nm, the smallest wavelength resolved, $\Delta\lambda$, would be:

$$\Delta\lambda = 300\, nm/62\,400$$
$$= 0.004\,81\, nm$$

Response 5.9

It can be seen from Figure 5.15 that the photomultiplier tube which gives the widest spectral wavelength coverage (190–900 nm) is one where the photocathode has a composition of Na–K–Sb–Ca, i.e. a trialkali.

Chapter 6

Response 6.1

The overall atomic weight of chlorine $= (35 \times 75.78/100) + (37 \times 24.2/100)$
$$= (26.50) + (8.95) = 35.45$$

Response 6.2

As the mass spectrometer is a vacuum system which operates from appropriate pumps (rotary, turbomolecular, etc.), it is the oil contained within the pumps which acts as a reservoir for the trace metal ions. Periodic replacement of the oil is therefore required.

Response 6.3

Sequential multi-element analysis is where one element at any one time is determined, although the solutions themselves may contain many other elements. In simultaneous multi-element analysis, all elements in a multi-element solution can be determined at the same time. In ICP–optical spectroscopy, sequential multi-element analysis would be carried out by using a monochromator (in the Czerny–Turner configuration), whereas simultaneous multi-element analysis would be carried out by using a polychromator (Paschen–Runge configuration).

Response 6.4

For this, the reader should consult Sections 6.5 and 5.4 and make appropriate notes for each of these.

Response 6.5

For zinc, an atomic mass of 66 is suitable (Zn, 27.8% abundant) in preference to an atomic mass of 64 at which Ni (0.95% abundant) occurs.

Response 6.6

For titanium, an atomic mass of 49 is suitable (Ti, 5.5% abundant) in preference to an atomic mass of 48 at which Ca (0.19% abundant) occurs.

Response 6.7

It appears that selenium (atomic masses of 78, 80 and 82) has an isobaric interference from krypton. As it is most unlikely that any krypton will be determined in the argon gas, this is not a problem. As we will see later, a major problem does occur at these masses (78, 80 and 82), but are of an 'alternative' type.

Response 6.8

Ba^{2+} ions occur at half the mass of the *parent* singly charged ion. In order of their importance, when considering the magnitudes of their effects (largest first), the Ba^{2+} ions will occur at atomic masses of 69, 68, 67, 66 and 65, respectively.

Response 6.9

From Equation (6.16):

$$A = [(x B_2 m_1 / m_2) - B_1] / (z - zx/y)$$

and therefore:

$$A = [0.9521 \times (3.5 \times 10^{-6}\,g \times 0.9215) \times (207.977/205.974)$$

$$- (3.5 \times 10^{-6}\,g \times 0.0126)]/(0.5254 - 0.5254 \times 0.9521/2.1681)$$

$$= [0.9521 \times (3.225 \times 10^{-6}\,g) \times (1.0097) - (4.410 \times 10^{-8}\,g)]/0.2947$$

$$= [(3.100 \times 10^{-6}\,g) - (4.410 \times 10^{-8}\,g)]/0.2947$$

$$= (3.056 \times 10^{-6}\,g)/0.2947$$

i.e. 1.037×10^{-5} g in 105.0 ml of solution

$$= 9.876 \times 10^{-8}\,g\,ml^{-1}, \text{ or } 0.0988\,\mu g\,ml^{-1}$$

Therefore, we have a concentration of $98.8\,ng\,ml^{-1}$, with is very close to $100\,ng\,ml^{-1}$.

Response 6.10

Assignment of the peaks to various elements, over the range from 46 to 240 amu, is shown in the annotated spectra presented in Figure SAQ 6.10 (a–f).

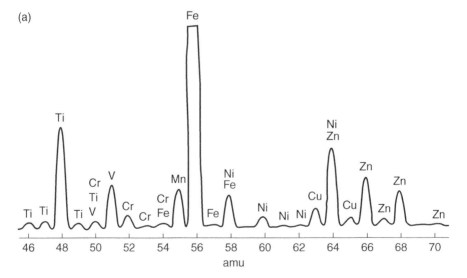

Figure SAQ 6.10 Annotated mass spectra showing the assignment of the peaks to the various elements [1]. From Dean, J. R., *Atomic Absorption and Plasma Spectroscopy*, 2nd Edition, ACOL Series, Wiley, Chichester, UK, 1997. © University of Greenwich, and reproduced by permission of the University of Greenwich (*continued*).

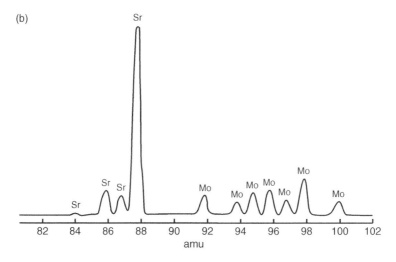

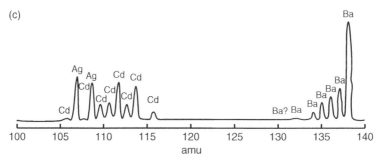

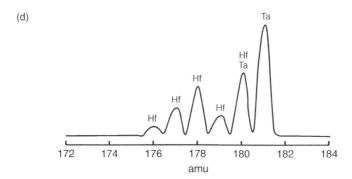

Figure SAQ 6.10 (*continued overleaf*).

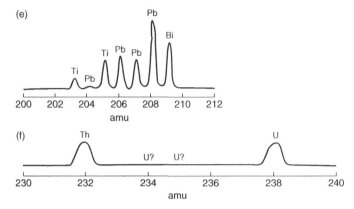

Figure SAQ 6.10 (*continued*).

Chapter 7

Response 7.1

The major component of paper is wood pulp – a natural material obtained from trees. As trees are grown in soil, its elemental composition will reflect its geographical origin. As each paper mill is likely to obtain its pulp from a variety of suppliers and locations, therefore the elemental composition will also vary.

Response 7.2

The various rare-earth elements are shown in the following table.

Symbol	Name
Sc	Scandium
Y	Yttrium
La	Lanthanum
Ce	Cerium
Pr	Praseodymium
Nd	Neodymium
Sm	Samarium
Eu	Europium
Tb	Terbium
Dy	Dysprosium
Ho	Holmium
Er	Erbium
Tm	Thulium
Yb	Ytterbium
Lu	Lutetium

Response 7.3

A dilution of $1 + 1999$ wt% was used in this case.

Response 7.4

A concentration of $10 \,\mu g \, l^{-1}$ was used in this case.

Response 7.5

The higher levels of the essential elements (Ca, Fe, Se and Zn) present in the milk whey samples may be due to their addition(s) during the production process.

Response 7.6

The presence of Al, Cd, Cr, Ni and Pb in the milk whey samples may result from the manufacturing processes involved in its production.

SI Units and Physical Constants

SI Units

The SI system of units is generally used throughout this book. It should be noted, however, that according to present practice, there are some exceptions to this, for example, wavenumber (cm^{-1}) and ionization energy (eV).

Base SI units and physical quantities

Quantity	Symbol	SI Unit	Symbol
length	l	metre	m
mass	m	kilogram	kg
time	t	second	s
electric current	I	ampere	A
thermodynamic temperature	T	kelvin	K
amount of substance	n	mole	mol
luminous intensity	I_v	candela	cd

Prefixes used for SI units

Factor	Prefix	Symbol
10^{21}	zetta	Z
10^{18}	exa	E
10^{15}	peta	P
10^{12}	tera	T
10^{9}	giga	G
10^{6}	mega	M

(continued overleaf)

Practical Inductively Coupled Plasma Spectroscopy   J. R. Dean
© 2005 John Wiley & Sons, Ltd

Prefixes used for SI units (*continued*)

Factor	Prefix	Symbol
10^3	kilo	k
10^2	hecto	h
10	deca	da
10^{-1}	deci	d
10^{-2}	centi	c
10^{-3}	milli	m
10^{-6}	micro	μ
10^{-9}	nano	n
10^{-12}	pico	p
10^{-15}	femto	f
10^{-18}	atto	a
10^{-21}	zepto	z

Derived SI units with special names and symbols

Physical quantity	SI unit		Expression in terms of base or derived SI units
	Name	Symbol	
frequency	hertz	Hz	$1\ Hz = 1\ s^{-1}$
force	newton	N	$1\ N = 1\ kg\,m\,s^{-2}$
pressure; stress	pascal	Pa	$1\ Pa = 1\ Nm^{-2}$
energy; work; quantity of heat	joule	J	$1\ J = 1\ Nm$
power	watt	W	$1\ W = 1\ J\,s^{-1}$
electric charge; quantity of electricity	coulomb	C	$1\ C = 1\ A\,s$
electric potential; potential difference; electromotive force; tension	volt	V	$1\ V = 1\ J\,C^{-1}$
electric capacitance	farad	F	$1\ F = 1\ C\,V^{-1}$
electric resistance	ohm	Ω	$1\ \Omega = 1\ V\,A^{-1}$
electric conductance	siemens	S	$1\ S = 1\ \Omega^{-1}$
magnetic flux; flux of magnetic induction	weber	Wb	$1\ Wb = 1\ V\,s$
magnetic flux density;	tesla	T	$1\ T = 1\ Wb\,m^{-2}$
magnetic induction inductance	henry	H	$1\ H = 1\ Wb\,A^{-1}$
Celsius temperature	degree Celsius	°C	$1°C = 1\ K$
luminous flux	lumen	lm	$1\ lm = 1\ cd\,sr$
illuminance	lux	lx	$1\ lx = 1\ lm\,m^{-2}$

Derived SI units with special names and symbols

Physical quantity	SI unit		Expression in terms of base or derived SI units
	Name	Symbol	
activity (of a radionuclide)	becquerel	Bq	$1\,Bq = 1\,s^{-1}$
absorbed dose; specific energy	gray	Gy	$1\,Gy = 1\,J\,kg^{-1}$
dose equivalent	sievert	Sv	$1\,Sv = 1\,J\,kg^{-1}$
plane angle	radian	rad	1^{a}
solid angle	steradian	sr	1^{a}

[a] rad and sr may be included or omitted in expressions for the derived units.

Physical Constants

Recommended values of selected physical constants[a]

Constant	Symbol	Value
acceleration of free fall (acceleration due to gravity)	g_n	$9.806\,65\,m\,s^{-2}$ [b]
atomic mass constant (unified atomic mass unit)	m_u	$1.660\,540\,2(10) \times 10^{-27}\,kg$
Avogadro constant	L, N_A	$6.022\,136\,7(36) \times 10^{23}\,mol^{-1}$
Boltzmann constant	k_B	$1.380\,658(12) \times 10^{-23}\,J\,K^{-1}$
electron specific charge (charge-to-mass ratio)	$-e/m_e$	$-1.758\,819 \times 10^{11}\,C\,kg^{-1}$
electron charge (elementary charge)	e	$1.602\,177\,33(49) \times 10^{-19}\,C$
Faraday constant	F	$9.648\,530\,9(29) \times 10^{4}\,C\,mol^{-1}$
ice-point temperature	T_{ice}	$273.15\,K$ [b]
molar gas constant	R	$8.314\,510(70)\,J\,K^{-1}\,mol^{-1}$
molar volume of ideal gas (at $273.15\,K$ and $101\,325\,Pa$)	V_m	$22.414\,10(19) \times 10^{-3}\,m^3\,mol^{-1}$
Planck constant	h	$6.626\,075\,5(40) \times 10^{-34}\,J\,s$
standard atmosphere	atm	$101\,325\,Pa$ [b]
speed of light in vacuum	c	$2.997\,924\,58 \times 10^{8}\,m\,s^{-1}$ [b]

[a] Data are presented in their full precision, although often no more than the first four or five significant digits are used; figures in parentheses represent the standard deviation uncertainty in the least significant digits.
[b] Exactly defined values.

The Periodic Table

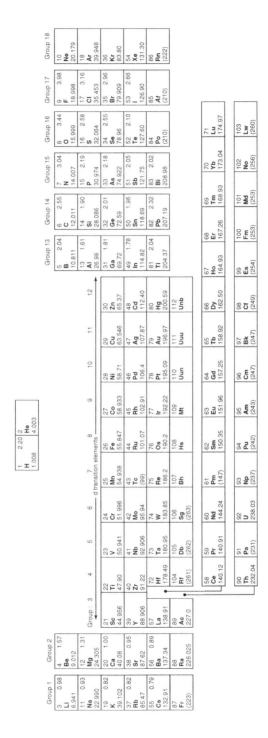

Group 1	Group 2												Group 13	Group 14	Group 15	Group 16	Group 17	Group 18

Key:

3   0.98 — Pauling electronegativity
— Atomic number
Li — Element
6.941 — Atomic weight (^{12}C)

1	2.20
H	
1.008	

2	
He	
4.003	

Group 1
3	0.98
Li	
6.941	
11	0.93
Na	
22.990	
19	0.82
K	
39.102	
37	0.82
Rb	
85.47	
55	0.79
Cs	
132.91	
87	
Fr	
(223)	

Group 2
4	1.57
Be	
9.012	
12	1.31
Mg	
24.305	
20	1.00
Ca	
40.08	
38	0.95
Sr	
87.62	
56	0.89
Ba	
137.34	
88	
Ra	
226.025	

d transition elements

Group: 3
21	44.956	**Sc**
39	88.906	**Y**
57	138.91	**La**
89	227.0	**Ac**

4
22	47.90	**Ti**
40	91.22	**Zr**
72	178.49	**Hf**
104	(261)	**Rf**

5
23	50.941	**V**
41	92.906	**Nb**
73	180.95	**Ta**
105	(262)	**Db**

6
24	51.996	**Cr**
42	95.94	**Mo**
74	183.85	**W**
106	(263)	**Sg**

7
25	54.938	**Mn**
43	(99)	**Tc**
75	186.2	**Re**
107		**Bh**

8
26	55.847	**Fe**
44	101.07	**Ru**
76	190.2	**Os**
108		**Hs**

9
27	58.933	**Co**
45	102.91	**Rh**
77	192.22	**Ir**
109		**Mt**

10
28	58.71	**Ni**
46	106.4	**Pd**
78	195.09	**Pt**
110		**Uun**

11
29	63.546	**Cu**
47	107.87	**Ag**
79	196.97	**Au**
111		**Uuu**

12
30	65.37	**Zn**
48	112.40	**Cd**
80	200.59	**Hg**
112		**Unb**

Group 13
5	2.04	**B**	10.811
13	1.61	**Al**	26.98
31	1.81	**Ga**	69.72
49	1.78	**In**	114.82
81	2.04	**Tl**	204.37

Group 14
6	2.55	**C**	12.011
14	1.90	**Si**	28.086
32	2.01	**Ge**	72.59
50	1.96	**Sn**	118.69
82	2.32	**Pb**	207.19

Group 15
7	3.04	**N**	14.007
15	2.19	**P**	30.974
33	2.18	**As**	74.922
51	2.05	**Sb**	121.75
83	2.02	**Bi**	208.98

Group 16
8	3.44	**O**	15.999
16	2.58	**S**	32.064
34	2.55	**Se**	78.96
52	2.10	**Te**	127.60
84		**Po**	(210)

Group 17
9	3.98	**F**	18.998
17	3.16	**Cl**	35.453
35	2.96	**Br**	79.909
53	2.66	**I**	126.90
85		**At**	(210)

Group 18
10		**Ne**	20.179
18		**Ar**	39.948
36		**Kr**	83.80
54		**Xe**	131.30
86		**Rn**	(222)

58	140.12	**Ce**
59	140.91	**Pr**
60	144.24	**Nd**
61	(147)	**Pm**
62	150.35	**Sm**
63	151.96	**Eu**
64	157.25	**Gd**
65	158.92	**Tb**
66	162.50	**Dy**
67	164.93	**Ho**
68	167.26	**Er**
69	168.93	**Tm**
70	173.04	**Yb**
71	174.97	**Lu**

90	232.04	**Th**
91	(231)	**Pa**
92	238.03	**U**
93	(237)	**Np**
94	(242)	**Pu**
95	(243)	**Am**
96	(247)	**Cm**
97	(247)	**Bk**
98	(249)	**Cf**
99	(254)	**Es**
100	(253)	**Fm**
101	(253)	**Md**
102	(256)	**No**
103	(260)	**Lw**

Index

Printed and bound by CPI Group (UK) Ltd, Croydon, CR0 4YY

27/10/2024